LSC $20.95
(anf)

P9-DFR-585

Master Math:

Solving Word Problems

$$\int = . \pm \sqrt{} < + \approx \div \times \neq \leq \pi \Delta \cong \Sigma >$$

ANALYZE ANY WORD PROBLEM,

TRANSLATE IT INTO

MATHEMATICAL TERMS, AND GET

THE RIGHT ANSWER!

Orillia Public Library
36 Mississaga St. W.
Orillia, Ontario
L3V 3A6

By

Brita Immergut

CAREER
PRESS
Franklin Lakes, NJ

Copyright © 2003 by Brita Immergut

All rights reserved under the Pan-American and International Copyright Conventions. This book may not be reproduced, in whole or in part, in any form or by any means electronic or mechanical, including photocopying, recording, or by any information storage and retrieval system now known or hereafter invented, without written permission from the publisher, The Career Press.

MASTER MATH: SOLVING WORD PROBLEMS
EDITED BY KRISTEN PARKES
TYPESET BY EILEEN DOW MUNSON
Cover design by The Visual Group
Printed in the U.S.A. by Book-mart Press

To order this title, please call toll-free 1-800-CAREER-1 (NJ and Canada: 201-848-0310) to order using VISA or MasterCard, or for further information on books from Career Press.

The Career Press, Inc., 3 Tice Road, PO Box 687,
Franklin Lakes, NJ 07417
www.careerpress.com

Library of Congress Cataloging-in-Publication Data

Immergut, Brita.
    Master math : solving word problems : analyze any word problem, translate it
into mathematical terms, and get the right answer! / by Brita Immergut.
        p. cm.
    Includes index.
    ISBN 1-56414-678-2 (pbk.)
        1. Word problems (Mathematics) 2. Problem solving. I. Title.

QA63.I45 2003
510—dc21

                                                                                2003053228

To my grandchildren
Michael, Jessica, and Kristina.
I hope you will always love mathematics.
Mormor

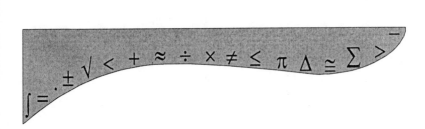

Table of Contents

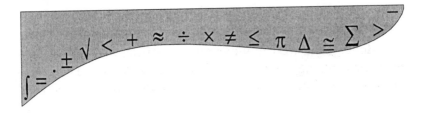

To the Reader

Many people are afraid of word problems. Why is that? Maybe it's because they remember that they had previous trouble with word problems. Or they think that they can't understand word problems because word problems are "difficult." Or they don't know how to unravel the problem to find out what the real question is. Or they simply don't know where to start.

In this book you will learn how to overcome those difficulties. You will be asked to read the problems slowly and to first concentrate on the *words* rather than on the *numbers*. Then you will learn how to break down the problem into smaller segments and to use a simple table to list the known numbers presented in the problem and the unknown number (usually x) that you are asked to figure out what it stands for. The solution for the problem usually involves the use of an equation and, for those of you who are a bit hazy about equations, you will find a short refresher in the Appendix.

The problems in this book mostly deal with situations from daily life: percents and discounts; interest (simple and compound); mixing of liquids and mixing of solids; ratios and proportions; and measurements in the English (customary) and the metric system and how to convert from one to the other.

There will also be problems dealing with the motion of cars, boats, and people at different speeds and how quickly work gets done. Then we will move on to statistics and probability problems: averages, graphs, probabilities, and odds. There will be rolling of dice, tossing of pennies, and drawing of playing cards.

Finally, you will learn how to solve word problems involving geometrical figures, such as triangles, polygons, circles, and cylinders. Some problems will deal with plane geometry, others with solid geometry, trigonometry, and analytic geometry. Each chapter contains not only worked-out problems, but also plenty of practice problems. The answers for the practice problems are at the end of the book.

I hope that when you are finished with this book you will feel as one of my former students did who told me: "Before I took your course I cried because I couldn't solve the word problems and now I cry because I am so happy that I can solve them."

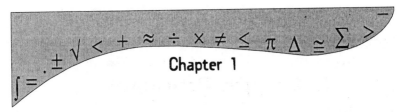

Chapter 1

Simple Equation Problems

In order to solve mathematical word problems we often need to use equations. In this chapter, you will learn how to set up simple equations to solve different kinds of word problems. For example, we will cut up a length of board or rope into shorter and longer pieces and, given the known total length and other facts, we will calculate the lengths of the pieces cut from it. In other examples we will calculate the ages of two children once we know how many years they are apart and what the sum of their ages is. We will also look at situations where one person weighs more or less than another and calculate each person's weight from the information given in the problem.

Then, we will learn the mathematical symbols for inequalities, that is, situations where something is greater than or smaller than something else and also how to solve problems in which we are told that something is at most so big or that something costs at least so much.

Finally, we will tackle word problems involving all kinds of numbers: positive and negative integers, including zero; odd and even integers; and consecutive integers.

The last example will show you how to solve a problem that requires the use of a quadratic equation.

(Note: If you want to brush up on your skills for solving equations, see the Appendix.)

Length Problems

Example:

Cut a 10-foot (ft.) long piece of wood into two pieces so that one piece is 2 ft. longer than the other. To solve this problem you have two choices:

By using algebra:

Call one piece x, then the other piece is $x + 2$.

Write an equation:

$$x + x + 2 = 10$$

Solve the equation:

$$2x = 8$$

$$x = 4$$

$$x + 2 = 6$$

$$\text{Total} = 10$$

Or by using arithmetic:

$10 - 2 = 8$ Take away the 2 ft. from the whole piece.

$8 \div 2 = 4$ Divide the piece by 2.

$4 + 2 = 6$ Add the 2 ft. to one of the pieces.

Total = 10

The pieces were 4 ft. and 6 ft.

Check your work by adding the pieces. Together they were 10 ft. Read the problem again to check all the facts.

Example:

A length of board was 10 inches shorter than another length. Together the boards were 20 inches. How long were the boards?

Call the long piece x and the short piece $x - 10$.

$$x + x - 10 = 20$$
$$2x = 30$$
$$x = 15$$
$$15 - 10 = 5$$

The pieces were 5 and 15 in.

Reread the problem. Is it true that the pieces equal 20 in. when put together? Is one piece 10 in. shorter than the other?

Practice Problems:

1.1   Solve the previous problem by calling the short piece of board x.

1.2   A 12-ft. rope is cut into three pieces so that the second piece is 1 ft. longer than the first and the third piece is 1 ft. longer than the second. How long are the pieces?

1.3   A 9-ft. board is cut into two pieces so that one piece is twice the other. How long are the pieces?

1.4   An 80-in. board is to be cut into three pieces so that one piece is twice another and the third piece is 10 in. more than the second. Find the length of each piece.

1.5   Two ropes are together 275 yards long. One rope is 50% longer than the other. How long are the ropes?

Age Problems

Example:

Leah is 2 years older than Tracy. Together the girls are 10 years old. How old are they?

Call Leah's age x. Then Tracy's age is $x - 2$.

$$x + x - 2 = 10$$
$$2x = 12$$
$$x = 6$$
$$x - 2 = 4 \qquad \text{Leah is 6 years old and Tracy is 4.}$$

Do you recognize this problem as essentially the same as the first length problem?

Practice Problems:

1.6   Elsa is 7 years younger than Thor. The sum of their ages is 35. How old are they?

1.7   Kristina's grandmother is 12 times as old as Kristina. Together they are 91 years. How old is the grandmother?

1.8   The sum of the ages of Jessica, her mother, and her grandmother is 100 years. The grandma is twice as old as the mother. Jessica's mother is 28 years older than Jessica. How old is grandma?

1.9   The sum of Eric's age and Lucas's age is 65. Two times Eric's age is the same as three times Lucas's age. Find the ages of the men.

1.10   Michael's age is multiplied by 7. Then 9 is added. The result is 93. How old is Michael?

Use of the Words "More Than" and "Less Than"

Beware of keywords! Many people believe that if they see "more than" in a word problem they must add and with "less than" they must subtract. That is not *always* the case. Look at the following examples:

Examples:

a) Sue weighs 5 pounds more than Amanda. If Amanda weighs 103 pounds, how much does Sue weigh?

b) Sue weighs 5 pounds less than Amanda. If Sue weighs 103 pounds, how much does Amanda weigh?

c) A school has an enrollment of 2381 students. This is 53 students less than last year. What was the enrollment last year?

d) A school had an enrollment of 2381 students last year. This is 53 students less than this year. What is the enrollment this year?

In each of these examples should you add or subtract?

Solutions:

a) Who weighs more? Sue. So we *add* 5 pounds to Amanda's 103 pounds.

b) Who weighs more? Amanda. So we *add* 5 pounds to Sue's 103 pounds

c) Is the enrollment larger this year? No, it is smaller by 53 students. We *add* 53 students to 2381.

d) Last year's enrollment was smaller than this year's. We *add* 53 to 2381.

Answers:

a) and b) 108 pounds; c) and d) 2434.

Practice Problems:

1.11 Six less than a number is 20. Find the number.

1.12 The sun rose at 7:45 a.m. That is 3 minutes earlier than yesterday. At what time did the sun rise yesterday?

1.13 The water level in the reservoir was 67.3 ft. That is 0.1 ft. higher than a month ago. What was the water level a month ago?

1.14 The population of the village is 487. That is 16 people less than 5 years ago. What was the population 5 years ago?

1.15 Joan weights 143 pounds. That is 10 pounds more than she weighed 5 years ago. How much did she weigh 5 years ago?

Inequalities Using the Words "At Least" and "At Most"

The symbols below translate into words the following way:

    $<$ is less than.

    $\leq$ is less than or equal to.

    $>$ is greater than.

    $\geq$ is equal to or greater than.

In word problems these symbols can also be used to translate:

    $\leq$ at most.

    $\geq$ at least.

Example:

Translate and solve:

1/3 of an unknown number is less than 4.

$$x/3 < 4$$

Multiply both sides by 3: $x < 12$

The unknown number can be any number smaller than 12.

Example:

If a number is subtracted from 10, the result is at most 22.

$$10 - x \leq 22$$

Subtract 10: $-x \leq 12$

Multiply by -1: $x \geq -12$

Note: When we multiply or divide an inequality by a negative number, we must reverse the inequality sign!

Answer: x can be any number that is equal to or greater than -12.

Practice Problems:

1.16 Compare -6 and -10 by using inequality symbols.

Solve the following inequalities:

1.17 The difference of a number and 7 is greater than or equal to 11.

1.18 Two times a number subtracted from 19 is less than or equal to 3.

Number Problems

Number problems deal with numbers, usually integers (whole numbers, which includes zero, and their negatives). An equation can be written once the problem has been stated. If the problem states something special about a number, that number is usually called x. Often these numbers deal with consecutive integers, which are integers that follow in order such as 5, 6, 7,... or -4, -3, -2,... Consecutive odd or even integers are two numbers apart, such as 2, 4, 6,... or 1, 3, 5,... If you add $1 + 3 + 5 = 9$, we have an odd sum of odd numbers, but $1 + 3 + 5 + 7 = 16$. A sum of four odd integers gives an even sum. Can a sum of even consecutive integers ever be odd?

Example:

The sum of three consecutive integers is 6. Find the integers.

Call the smallest integer x. Then the others are $x + 1$ and $x + 2$

The sum is $x + x + 1 + x + 2$ or 6.

$$3x + 3 = 6$$
$$3x = 3$$
$$x = 1$$
$$x + 1 = 2$$
$$x + 2 = 3$$

The consecutive integers are 1, 2, and 3.

Check: $1 + 2 + 3 = 6$

Example:

The sum of four consecutive even integers is 44. Find the integers.

Call the integers x, $x + 2$, $x + 4$, $x + 6$.

The sum is $x + x + 2 + x + 4 + x + 6$ or 44.

$$4x + 12 = 44$$
$$4x = 32$$
$$x = 8$$
$$x + 2 = 10$$
$$x + 4 = 12$$
$$x + 6 = 14$$

The four integers are 8, 10, 12, and 14.

Check: $8 + 10 + 12 + 14 = 44$

Practice Problems:

1.19 Find five consecutive odd integers whose sum is 55.

1.20 The sum of three consecutive even integers is –12. Find the integers.

1.21 The sum of five consecutive integers is 0. Find the integers.

1.22 The sum of four consecutive even integers is 2 more than five times the first integer. Find the first integer.

1.23 Two consecutive even integers have a sum of 14. What is their product?

Some types of number problems may sound more complicated, but they can also be solved by simple equations.

Example:

Find three consecutive odd integers so that four times the first integer equals the sum of the other two.

Call the smallest integer x. The others are $x + 2$ and $x + 4$.

$$4x = x + 2 + x + 4$$
$$4x = 2x + 6$$
$$2x = 6$$
$$x = 3$$
$$x + 2 = 5$$
$$x + 4 = 7$$

The odd integers are: 3, 5, and 7.

Check: $4(3) = 12$ $\qquad\qquad 5 + 7 = 12$

Practice Problems:

1.24 Find the largest of three consecutive even integers when six times the first integer is equal to five times the middle integer.

1.25 The sum of four consecutive integers is 186. Find the integers.

1.26 Find three consecutive even integers such that three times the first equals the sum of the other two.

1.27 Five times an odd integer plus three times the next odd integer equals 62. Find the first odd integer.

1.28 Find three even integers, if five times the first integer, plus twice the second integer, plus the third integer equals 40.

The following example and practice problems require the knowledge of how to solve quadratic equations. If you have never done this, you might want to skip them. However, they are also explained in the Appendix. All of the following problems can be solved by factoring.

Example:

The product of two consecutive integers is 20. Find the integers.

Call the integers x and $x + 1$.

The product is $x(x + 1) = 20$

$$x^2 + x = 20$$

$$x^2 + x - 20 = 0$$

We factor this equation by finding two integers whose product is -20 and whose sum is 1 (the coefficient of x). The integers are -4 and 5.

$$(x - 4)(x + 5) = 0 \qquad x - 4 = 0 \qquad x + 5 = 0$$
$$x = 4 \qquad\quad x = -5$$
$$x + 1 = 5 \qquad x + 1 = -4$$

The consecutive integers are 4 and 5, or –5 and –4.

Note: Quadratic equations always have two possible solutions.

Practice Problems:

1.29 The product of two consecutive positive integers is 20. Find the integers.

1.30 One number is 4 more than another number. Their product is 5. Find the numbers.

1.31 The product of two consecutive odd integers is 15. Find the integers.

1.32 The product of two numbers is 243. One number is three times the other one. Find the numbers.

1.33 One number is 2 more than another. Their product is 440. Find the numbers.

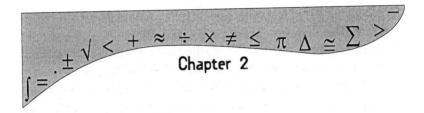

Chapter 2

Percents

In this chapter you will learn how to solve word problems that deal with percent (%). In daily life, percents are used all the time by banks, in statistics, in advertising, in store discounts, and in many other situations.

We will start with problems involving conversions from percents to decimals and fractions, and from decimals and fractions to percents. We also want to calculate percents of numbers (for example, what is 30% of 90) and express the answer in fractions or decimals.

Then we will solve problems for situations where the percent is already included in the final number. This happens in grocery and other stores where the sales tax (which is a percentage of the sales price) is included in the total amount of money you pay. Similarly, some restaurants include a 15% tip in the total charge.

You will also learn how to calculate percent increases and decreases. For example, when the subway or bus fare is increased by 50¢ from $1.50 to $2.00, what is the percentage increase? Or when postage stamps are increased from 34¢ to 37¢?

We will then go on to discounts, and discounts on already discounted items.

This chapter also deals with interest calculations: interest paid by banks on your deposit in a savings account and interest charged by banks and credit card issuers on money you borrowed or charges on your credit card. The difference between simple interest and compound interest will also be made clear.

Finally, we will learn about investments in stocks and bonds and how to calculate gains or losses from these investments. We will give examples of word problems dealing with other situations of profit and loss.

Percents and Decimals

The word percent consists of two parts: per and cent. *Per* means "divide by" and *cent* means "hundred."

Example:

a) $100\% = 100 \div 100 = 1$ (that is, "the whole thing")

b) $50\% = 50 \div 100 = \dfrac{50}{100} = \dfrac{1}{2}$ or 0.5

c) $1/2\% = \dfrac{1}{2} \div 100 = \dfrac{1}{2} \times \dfrac{1}{100} = \dfrac{1}{200}$ or 0.005

d) $0.25\% = 0.0025$

When a number in decimal form is divided by 100, the decimal point is moved two steps to the left: $100.0\% = 1.00$, $50.0\% = 0.5$, $5.0\% = 0.05$, etc.

Practice Problems:

2.1   Change to a fraction: $1/3\%$

2.2   Change to a fraction: $4\ 1/2\%$

2.3   Change to a decimal: 0.006%

2.4 Change to a decimal: 1.7%

2.5 Change to a whole number: 400%

Often, we have to take some percent *of* a number. That means multiply the percent by the number.

Example:

Find:

    a) 30% of 60

    b) 24% of 300

    c) $33\frac{1}{3}$% of 30

Solution:

    a) $0.3 \times 60 = 18$

    b) $0.24 \times 300 = 72$

    c) $33\frac{1}{3}\% \times 30 = \frac{100}{3} \times \frac{1}{100} \times 30 = 10$

Practice Problems:

2.6 Find 50% of 600.

2.7 Find 2.5% of 7.50.

2.8 Find 3.5% of 15,000.

2.9 Find 2/3% of 60.

2.10 Find 15% of 80%.

Be careful with calculators! They are very useful, but you must know how they work. For example, in math terms,

$$100 + 10\% - 10\% \text{ means } 100 + \frac{10}{100} - \frac{10}{100} = 100$$

However, many calculators "read" 10% as 10% *of* the number before it. "Of" means multiplication, so

$$100 + 10\% \text{ of } 100 = 100 + \frac{10}{100} \times 100 = 110 \text{ then}$$

$$110 - 10\% \text{ of } 110 = 110 - \frac{10}{100} \times 110 = 99.$$

Sometimes you will need to translate a number to a percent, that is, you need to insert the percent symbol. Remember that 100% equals 1. You probably also know that you can multiply any number by 1 and not change the number. We use that property to change a number to a percent.

Example:

Change the following numbers to percents:

   a) 2   b) 0.7   c) 3/4

   Solution:

   a) 2 × 100% = 200%

   b) 0.7 × 100% = 70%

   c) 3/4 × 100% = 300/4% = 75%

Practice Problems:

Change to percent:

   2.11 0.0056

   2.12 12

   2.13 4/5

   2.14 2/3 (Use common fractions or decimals to two placed rounded off.)

   2.15 500

Percents and Numbers

These problems come in three forms:

1. What percent of a number is another number?

2. A known % of what number is another number?

3. A known % of a known number is what number?

Using algebra, we can write three equations where x is the unknown ("what") number and a and b are known numbers:

1. $x\% \cdot a = b$ (In math problems, multiplication is shown by a dot between the numbers that are to be multiplied.)

2. $a\% \cdot x = b$

3. $a\% \cdot b = x$

Example:

Solution:

1. What % of 60 is 15?

$$1. \quad x\% \cdot 60 = 15$$

$$\frac{x}{100} \cdot 60 = 15$$

$$x = \frac{1500}{60}$$

$$x = 25$$

2. 25% of what number is 15?

$$2. \quad 25\% \cdot x = 15$$

$$\frac{25}{100} \cdot x = 15$$

$$x = \frac{1500}{25}$$

$$x = 60$$

3. 25% of 60 is what?

$$3. \quad 0.25 \cdot 60 = 15$$

Shortcut:

1. $\dfrac{15}{60\%} = 25$

2. $\dfrac{15}{25\%} = 60$

3. No shortcut is possible.

Practice Problems:

2.16 13 is what % of 52?

2.17 10% of what number is 635?

2.18 The enrollment at a local college is 3500. Of these, 30% are math majors. How many math majors are there at the college?

2.19 Suppose 120 students out of 150 passed a math course. What percent is that?

2.20 Lisa spends $175, or 25%, of her monthly take-home pay for rent. What is her monthly salary?

The Percent Is Included

Example:

What number increased by 6% is 2544?

       100% + 6% = 106%

       106% of what number is 2544?

$$\dfrac{2544}{1.06} = 2400$$

Check: 6% of 2400 = 144 and 2400 + 144 = 2544

If you prefer, you can use an equation:
$$x + 6\% \cdot x = 2544$$
$$x + 0.06x = 2544$$
$$1.06x = 2544$$
$$x = \frac{2544}{1.06}$$
$$x = 2400$$

Example:

A merchant counted the money in his cash register and found that he had $4320. A sales tax of 8% was included. How much money did the merchant keep and how much did he owe the government?

Let's analyze the problem:

If his money is x, the tax rate is 8%, and he counted out $4320, we have:
$$x + 8\% \cdot x = 4320$$
$$1.08x = 4320$$
$$x = \frac{4320}{1.08}$$
$$x = 4000$$

Therefore, the merchant keeps $4000 and the government gets $320.

Shortcut:

Do the division directly without using x: $\dfrac{4320}{1.08}$

Example:

a) Eva paid $23 for a taxi ride with 15% tip included. How much did the trip itself cost?

b) $5.40 has been collected for an item with 8% sales tax included. How much did the item cost before the tax was included?

Solutions:

a) $\dfrac{23}{1.15} = 20$ Answer: $20.00

b) $\dfrac{5.40}{1.08} = 5$ Answer: $5.00

Practice Problems:

2.21 Bailey paid the bank $127.20 every month to repay principal on her loan plus 6% interest. How much was the principal repayment?

2.22 What number increased by 6% of itself is 371?

2.23 An item cost $82.40 including a 3% tax. What was the cost of the item itself?

2.24 After a salary increase of 4%, Nell's salary was $3120/month. What was her salary before the increase?

2.25 The price of a pound of prime steak was increased by 8% to $12.42. What was the price before the increase?

Percent Increase and Decrease

Example:

In many problems we have information "before and after." For example, the price of eggs was $1.29 for a dozen eggs last month and now it is $1.49. What was the percent increase?

The original price of the eggs was $1.29 and the increase was $1.49 − $1.29 = $0.20. The problem can then be restated: What % of 1.29 is 0.20?

Equation: $x\% \cdot 1.29 = 0.20$

$$x = 15.5 \text{ (rounded)}$$

$$\frac{.20}{1.29} \times 100$$

The price increased by 15.5%.

Example:

Let's change the problem and pretend that the eggs decreased from $1.49 to $1.29. What was the percent decrease?

Now the original price is $1.49 and the decrease was $0.20.

$$x\% \cdot 1.49 = 0.20$$

$$x = 13.4 \text{ (rounded)}$$

The price decreased by 13.4%.

Practice Problems:

2.26 Jill's monthly salary as a teacher increased from $2600 to $2675. What was the percent increase?

2.27 During one year the price of a certain stock decreased from $75.6 a share to $37.5. What was the percent decrease (rounded to the nearest percent)?

2.28 Ellen paid $29,000 for a new car. After a year, she found that the value of the car was only $24,500. Find the percent decrease.

2.29 The enrollment at a community college increased from 6783 to 7895. What was the percent increase?

2.30 The price of a house increased 6% in one year. What is the current value of the house, if it was worth $250,000 before the increase?

Discounts

Stores often advertise a reduction of their prices with words such as "off" or "discounts."

Example:

The price of a dress was reduced from $200 to $180. What was the percent of discount?

> The discount was $20 and the original price $200. The problem can be rewritten as "What percent of 200 is 20?"

$$x\% \times 200 = 20$$

$$x = 10$$

The percent of discount is 10%.

Practice Problems:

2.31   A bookstore gives a 10% discount for students. What does a student pay for a book that originally cost $35?

2.32   In a certain store, a customer is given a $12.80 discount for a total order of $256. What is the percent of discount?

2.33   A customer is getting a discount of $13.80 for a purchase of $276. What is the percent of the discount?

2.34   A jacket that used to cost $150 is for sale with a discount of 15%. How much is the jacket now?

2.35   A price club gave a 15% discount on all items on a certain day. What do you pay for an item that usually costs $236?

Example:

Jane was offered a 10% discount on a purchase and pays $22.50. How much did the item cost before the discount?

$100\% - 10\% = 90\%$

Jane paid 90% of the original price.

$$90\%x = 22.50$$
$$x = \frac{22.50}{90\%}$$
$$x = 25$$

Answer: $25.00

Shortcut: Do the problem without using x (that is, divide 22.5 by 0.9).

Example:

a) The price of an elegant winter coat with a 20% end-of-the-season markdown is $840. What was the price before the markdown?

b) A discount of 20% gives a sale price of $94. What was the original price?

c) Gwen bought a dress on sale for $62.50. It was on sale for 75% off. What was the original price?

Solutions:

a) $\dfrac{840}{80\%} = 1050$ Answer: $1050

b) $\dfrac{94}{.8} = 117.5$ Answer: $117.50

c) $\dfrac{62.5}{25\%} = 250$ Answer: $250

Practice Problems:

2.36 A pair of shoes was discounted with 16% and sold for $84. What did the shoes cost originally?

2.37 A lady bought a coat marked 20% off for $144. How much had the coat cost originally?

2.38 The price of a sweater was lowered from $75 to $63.50. What was the discount in percent?

2.39 Certain movie theaters give discounts of 15% to senior citizens. If Joe paid $5.10 for his ticket, how much did he save?

2.40 Millie was thinking of buying a microwave oven. She saw an ad: Discount 35% (you save $70). Another ad stated: Discount 45%, (you save $90). Which oven is cheaper?

Discounts on Discounts

Stores sometimes offer discounts on already discounted items. For example, an item is offered at a 15% discount, and then is advertised with an additional 40% discount. Do you get a discount of 55%?

Let's analyze the problem:

The 15% discount lets you pay 100% − 15% = 85% of the price. Then you get a discount of 40% of 85% of the price, which is 0.4 × 85% = 34% of the price. The total discount is 15% + 34% = 49%.

An alternative to solving this problem is: The first discount lets you pay 85% of the original price. The second discount lets you pay 100 − 40% = 60% of the new price. That is 60% of 85%, which is 51% of the original price. If you pay 51%, you get a 49% discount.

Example:

Macy's offered a super sales day with 40% off certain already discounted items. What is your actual discount rate if the item already had a discount of *a*) 10%, *b*) 25%, *c*) 34%?

Solutions:

a) The first discount is 10%. Then the price of the item is 90% of the original price. You get a 40% discount of that price, which is 40% of 90% or 36% of the original price.
The total percent discount is 10% + 36% = 46%.

Alternate solution:
(100% – 10%) = 90%
The new price is 90% of the original price.

(100% – 40%)90% = 60%(90%) = 54%
The price is 60% of 90% of the original price. You pay 54% after two discounts.

100% – 54% = 46%   The discount is 46%.

b) The original discount was 25%. The new price was 75% of the original price. The second discount was 40% of 75% = 30%.
The total discount is 25% + 30% = 55%.

Alternate solution:
(100% – 25%) = 75%
The new price is 75% of the original price.

(60%)(75%) = 45%
You pay 45% after the second discount.

100% – 45% = 55%   The total discount is 55%.

c) The original discount was 34%. The new price is 66% of the original price. The second discount is 40% of 66% = 26.4%.
The total discount is 34% + 26.4% = 60.4%.

Alternate solution:

$(100\% - 34\%) = 66\%$

The new price is 66% of the original price.

$(60\%)(66\%) = 39.6\%$

You pay 39.6% after two discounts.

$100\% - 39.6\% = 60.4\%$ The total discount is 60.4%.

Practice Problems:

2.41 How much do you pay for an item that originally cost $100, if you get two discounts, one of 30% and one of 40%? Does it matter which discount you use first?

2.42 An item cost $240. The price was lowered twice, first with 15% and later with 25%. What was the final price?

2.43 A radio cost $120. The price was lowered first by 15% and later by 12%. How much was it then?

2.44 A washing machine cost $860. The price was first lowered by 35% and later by $68. With how many percent was the price lowered the second time?

2.45 A company has 540 employees. How many employees will the company have after 2 years if the number of employees increases by 10% each year?

Interest

When you have a savings account in a bank, you earn interest. If you borrow money, you pay interest.

Interest can be *simple* or *compound*. With simple interest we multiply the *principal* (the money we deposit or borrow) by the interest *rate* and the *time* the money is in the bank or used by us. The interest rate is the percentage per year and the time is in years. Unfortunately, for our calculations, simple interest is not often used, but it is a starting point for understanding the procedure of calculating interest.

Simple Interest

Formula:

I(nterest) = P(rincipal) × r(ate) × t(ime) or *I = Prt*

Example:

Find the interest if $1000 is deposited in the bank at 3% simple interest for five years.

$I = 1000 \times 3\% \times 5 = 1000 \times 0.03 \times 5 = 150$

The interest is $150.

Example:

Find the interest rate if $1000 is kept in the bank for five years and gives a simple interest of $150.

Use the formula $I = Prt$: $150 = 1000\, r(5)$.

This is an equation; to solve for r, divide both sides by 5000.

$\dfrac{150}{5000} = r$ or $r = 0.03 = 3\%$

Practice Problems:

2.46 Use the interest formula $I = Prt$ to find:

a) r when $P = \$7500, I = \$600$, and $t = 4$ years.

b) t when $P = \$4000, I = \$400$, and $r = 5\%$.

2.47 If you borrow $600 for a year and repay the whole loan at the end of the year with $636 including simple interest, what rate of interest are you paying?

2.48 Ed opens an insured money market account with $500. How much money will he have after 6 months if the interest rate is 0.75% per year?

2.49 If Ed invests his $500 in a CD (Certificate of Deposit) account instead, which pays 1.14% per year, how much money will he have after 6 months?

2.50 Fred borrowed $20,000 for 5 months at a 16% annual interest rate in order to buy a new car. How much money will he have to pay back?

Credit Cards

You receive a credit card when a lending institution, such as a bank, offers to lend you money. Department stores, gasoline companies, and other institutions also offer credit cards. When you make a purchase using a credit card, the lender pays the seller for the item you bought less a small commission (2–3%) and then sends you a bill at the end of the month. The bill usually lists the date of payment to the seller, the name of the seller, and the amount. You usually have a choice to pay the entire bill by a certain listed date or to pay a minimum amount and carry over the balance owed to your next monthly bill. The statement from the credit card company also shows how they determine your finance charge. This is the interest you owe your lender for the money they have loaned to you by paying the seller for the item you purchased with your credit card.

Example:

On March 4, you bought books at a bookstore for $21.60. You also had the *New York Times* delivered to your home 5 days a week for $19 billed directly to your credit card on March 9. Total charges from February 21 to March 21 (statement closing date) were $40.60. This is the new balance on your credit card. If you pay by the payment due date of April 15, there will be no finance charge (that is, no interest to be paid to the lender).

The statement also lists a minimum payment due of $10 to be paid by the payment due date. If the lender charges you interest at an annual percentage rate (APR) of 10.15%, what will be the finance charge on the remaining balance of $40.60 – $10 = $30.60 on your next bill of April 21, assuming you purchased no additional items before April 21?

Solution: An annual percentage rate of 10.15% / 365 days = a daily rate of 0.02781%.

The lender lent you
$21.60 from March 4 to March 21 = 18 days
and $19 from March 9 to March 21 = 14 days.

$21.60 × 18 × 0.02781/100 = $0.1081 or 10.81¢

$19 × 14 × 0.02781/100 = $0.07397 or 7.4¢

Total finance charge = 18¢.

Note: If you purchased additional items between the last closing statement (March 21) and the next one (April 21) then interest will be charged on the average daily balance, which includes the additional items.

Practice Problem:

2.51 You have purchased a computer for $1000 on March 10. The minimum balance was $50 due on March 21. The APR is 9.5%. What will be the finance charges for the remaining balance on your next statement on April 21, assuming you made no other purchases until then?

Compound Interest

Banks usually use compound interest to calculate the earnings on our money. For example, if you deposit $100 at a yearly rate of 2% and it is compounded (interest is added to the principal) once a year, you have:

After 1 year: $100 + 2%$ of $100 = $102

After 2 years: $102 + 2%$ of $102 = $104.04

After 3 years: $104.04 + 2%$ of $104.04 = $106.12

If your money is compounded daily, it is too cumbersome to calculate the interest step by step. We need a formula!

The accumulated value A is the principal plus interest $P + Prt$, where P is the principal, r the rate per year, and t the time (in years) for one compounding period, so

After 1 year, the accumulated value is:
$$P + Prt = P(1 + rt)$$

After 2 years, the accumulated value is:
$$P(1 + rt) + P(1 + rt)rt = P(1 + rt)(1 + rt)$$
$$= P(1 + rt)^2$$

After 3 years, the accumulated value is: $P(1 + rt)^3$

The general formula is $A = P(1 + rt)^n$, where n is the number of compounding periods. In our case $P = 100, r = 2%$ and $t = 1$, and n $= 3$, so $A = 100(1 + 2% \cdot 1)^3 = 100(1.02)^3 =$ (Use your calculator!) 106.12.

In order to calculate the power with a calculator, do the following:
$$1.02, y^x, \times, 100 =$$

Example:

If $1000 were compounded monthly for 3 months at an annual interest of 10%, how much interest would there be?

Formula: $A = P(1 + rt)^n$,
where $P = $1000, r = 10% \div 12, t = 1$, and $n = 3$

$A = 1000(1 + 10% \div 12)^3 = 1000(1.00833)^3$
$\quad = 1,025.21$ (rounded)

$I = 1.025.21 - 1000 = 25.21$

The interest would be $25.21.

The simple interest would have been:
1000 × 10%/12 × 3 = 25 or $25

Practice Problems:

2.52 Redo the example without the formula.

2.53 Calculate the interest on $500 for one year at 2% compounded monthly.

2.54 Find the new principal if $700 gets a 1.5% interest compounded daily for one year. Use 365 for the number of times the money is compounded per year.

2.55 Barbara borrows $12,000 with an interest of 12% compounded yearly. How much interest does she owe after 4 years?

2.56 Viviane bought treasury notes for $6800. They are supposed to be worth $10,000 after five years. What is the percent interest if the money is compounded every year? Is it 6%, 7%, 8%, or 9%?

Hint: Use your calculator to test the different percents.

Bank Deposits

Example:

Leslie deposits part of $3000 in a certificate of deposit (COD) that pays simple annual interest of 2.71% and the rest in a passbook savings account that pays 0.75% compounded annually. If she earns $34.26 in interest in one year totally, how much did she deposit in her savings account?

Assume she deposited x dollars in her savings account. Then she must deposit $3000 - x$ dollars in her CD.

Table:

Investment	Percent	Interest
x	0.75%	$0.0075x$
$3000 - x$	2.71%	$(3000 - x)0.0271$
Total 3000		34.26

$$0.0075x + (3000 - x)0.0271 = 34.26$$
$$0.0075x + 81.3 - 0.0271x = 34.26$$
$$81.3 - 34.26 = 0.0271x - 0.0075x$$
$$47.04 = 0.0196x$$
$$x = 2400$$

Leslie deposited $2400 in her savings account.

Practice Problem:

2.57 One bank pays 0.50% on day-to-day savings
      compounded once a year while the credit union
      pays 2.72% compounded quarterly. If you had
      $1500 to invest for three years, how much more
      would you earn if you put your money in the credit
      union?

Investments

Whether you put money in the bank, in the stock market,
or in bonds, you *invest* your money. What you earn on your
investments differs. With CDs, as well as with a savings account,
you know your interest rate. With stocks and bonds dividends
can vary. However, the calculations we can make about our
accounts are all similar.

Stocks

Stocks are shares in a corporation. The price of each share will vary according to the perceived value of the share by investors. When many investors want to buy shares of a certain company, the price per share goes up. When many investors want to sell shares of a company, the price per share goes down. In order to reward shareholders of profitable companies, the company will declare a dividend, usually on a quarterly basis.

Example:

Joe buys 100 shares of Company A at a price of $10/share on January 2. The stock pays a quarterly dividend of $1.50. Joe sells his shares on January 2 of the next year at a price of $12/share. What is his total return on his investment *a*) in cash and *b*) in percent?

Joe paid 100 × $10 = $1000.
He received 100 × $12 = $1200.

Profit = $200, but in addition he received 4 quarterly dividends of $1.50 = $6. So, his total return on his investment of $1000 was $206 or 20.6%.

Practice Problem:

2.58   Jim bought 100 shares of Company B at a price of $10/share on January 2. The stock pays a quarterly dividend of $1.25 and sold on January 2 the next year for $12.05/share. What was Jim's total profit?

Bonds

Bonds are promissory notes issued by a corporation, by the federal government, by a city, or any other entity in order to raise money to pay its expenses or debts.

Bonds are bought and sold in multiples of 10.

Bonds have a "maturity date," that is, the year and date when the issuer of the bond has to pay the bond holder the "face value" of the bond. The face value of a bond is $100, but its actual price will vary according to its quality or "rating," that is, to what extent bond analysts trust in the ability of the bond issuer to repay the buyer the face value at maturity, as well as the interest the issuer pays to the bond holder. An important quantity to consider when buying or selling a bond is its "yield," which is the interest divided by the current price of the bond. The yield will increase when the price decreases and the yield will decrease when the price increases. The investor must also consider whether to buy a tax-free municipal bond or some company bond from which the interest is not tax-free. There are also junk bonds, which are high-risk (not very safe!) bonds but, accordingly, pay the buyer a higher interest.

Example:

On January 2, 2003, Mary bought 10 bonds maturing in 2010 and paying 5% interest. She paid $95 for each bond, for a total of $950. She also bought 10 bonds maturing in 2020 paying 6% interest for $90 each, a total of $900. If she keeps all 20 bonds to maturity:

a) How much cash will she receive for the bonds?

b) How much will she have earned in interest by the end of 2010 and by the end of 2020?

c) What is the yield of the 2010 bonds in 2003 and in 2010? What is the yield of the 2020 bonds in 2003 and in 2020?

Solutions:

a) If she keeps all bonds to maturity, she will receive the face value for all of them. In 2010 she will receive $10 \times \$100 = \$1000$, in 2020 she will receive $10 \times \$100 = \$1000$.

b) For the 2010 bonds, by the end of 2010, Mary will have earned 8 (years) × 10 (bonds) × $5 (interest per bond per year) = $400.

For the 2020 bonds, by the end of 2010, she will have received 8 × 10 × $6 = $480, and by the end of 2020 she will have received an additional 10 × 10 × $6 = $600. So by the end of 2010 she will have earned $880 and by the end of 2020 she will have earned a total of $1480.

c) The yield of the 2010 bonds in 2003 is 5% divided by $95 = 5.2%; in 2020 the yield will be down to 5% because the price of the bond will be $100 (face value). For the 2020 bonds in 2003, the yield is 6% divided by $90 = 6.6% and in 2020 it will be down to 6% because the price of these bonds is now $100 (face value).

Practice Problem:

2.59 Jane took a risk on January 2, 2003, and bought 10 "junk bonds" paying interest at 15% and maturing in 2010, at a price of $50.

a) Assuming that the bond issuer is still in business by 2010, how much interest will Jane have collected from 2003 to 2010?

b) How much will Jane collect from the bond issuer when she cashes in her bonds on maturity?

c) What was the yield of these bonds at the time Jane bought them?

Profit and Loss

We have already discussed profit when it comes to stocks. But store owners often use a formula to calculate their profit or loss on different items. A loss would be a negative profit.

Example:

A grocery store owner sold milk for 99¢/quart. Each quart cost him 80¢ and his operation costs were 15% of the cost price. What was his profit or loss on the milk?

15% of 80 cents = 12¢

99¢ – 80¢ – 12¢ = 7¢

His profit was 7¢ per quart.

If instead his operation costs were 25% of the cost price, his profit would be:

99¢ – 80¢ – 25%(80)¢ = (99 – 80 – 20)¢ = –1¢

He would have a loss of 1¢ per quart.

Practice Problems:

2.60 A glass vase cost the owner of a store $50. The markup was 40%. How much profit did the owner get if the vase was sold at 15% off the advertised selling price?

2.61 If an item is sold for $60, there is a profit of 20%. If the same item is sold at a loss of 20%, what is the selling price?

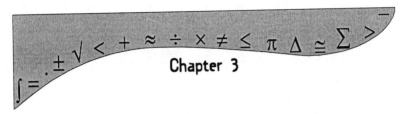

Chapter 3

Advanced Level Age Problems

In this chapter we will continue to analyze and solve age-related word problems, but the ones in the following pages are more complicated than the ones we solved previously. Therefore, we will have to spend more time to analyze and understand what the question is and how we can set up the equation, or sometimes two equations, that will help us to come up with an answer.

As before, these problems involve age differences between one child and another, between a child and an adult, or more than two persons, and we will have to read the problems carefully in order to calculate the ages of these persons now, or in a given number of years.

By setting up tables that list the facts and then generating equations from these tables , we will be able to solve the problems and come up with the correct answers. At the end of the chapter we will learn how to set up quadratic equations needed to solve certain kinds of problems.

Example:

Alan is 5 years older than Bert. If, in 3 years, Alan's age will be twice Bert's, how old are Alan and Bert now?

If Bert's age is x, then Alan's age is $x + 5$. Set up a table:

Name	Age now	Age in 3 years
Alan	$x + 5$	$x + 5 + 3 = x + 8$
Bert	x	$x + 3$

Alan's age is twice (2 times) Bert's.

Equation: $x + 8 = 2(x + 3)$

$$x + 8 = 2x + 6$$

$$2 = x$$

Bert's age is 2 and Alan's age is $2 + 5 = 7$

Check: In 3 years $2 + 3 = 5$

$$7 + 3 = 10$$

$$2 \times 5 = 10$$

This problem could also have been solved with Alan's age as x, but keep in mind that if Alan is older than Bert, then Bert is younger than Alan. Therefore:

Name	Age now	Age in 3 years
Alan	x	$x + 3$
Bert	$x - 5$	$x - 5 + 3 = x - 2$

Equation: $x + 3 = 2(x - 2)$

$$x + 3 = 2x - 4$$

$$7 = x$$

Alan is 7 and Bert is $7 - 5 = 2$.

Example:

Elsa is 7 years younger than Thor. The sum of their ages is 35. How old are they?

If Thor is x, then Elsa is $x - 7$.
Together they are $x + x - 7$ or 35.

Equation:
$$x + x - 7 = 35$$
$$2x = 35 + 7$$
$$2x = 42$$
$$x = 21$$

Thor is 21 and Elsa is 14.

Practice Problems:

3.1 Do the previous exercise with Elsa's age as x.

3.2 When I was born, my father was 41 years old. Eight years ago he was 3 times as old as I will be in 5 years. How old am I?

3.3 Lyn's father is 4 times as old as Lyn. Five years ago the father was 7 times as old as Lyn. How old is Lyn now?

3.4 Ed and Carl are brothers. The sum of their ages and their father's age is 61 years. Ed is 5 years older than Carl and their father is 6 times as old as Carl. How old is Carl?

3.5 A 40-year-old man has three daughters, ages 6, 3, and 1. In how many years will the combined ages of his daughters equal 80% of his age?

Example:

The sum of Tom's and Lou's age is 41 and the difference is 31. How old are they?

Here we work with both x and y. Let x stand for Tom's age and y stand for Lou's age.

Equations: $x + y = 41$

$$x - y = 31$$

Add the equations: $2x = 72$

$$x = 36$$

Tom is 36 years old and Lou is $41 - 36 = 5$ years old.

Now, if the problem were stated differently: The sum of Lou's age and his father's is 41 and the difference is 31, we must realize that the father is older. If we call Lou's age x and the father's age y, we get:

$$x + y = 41$$

and $y - x = 31$

$$2y = 72$$

$$y = 36$$

The father is 36 years old and Lou is 5 years old.

Practice Problems:

Use two variables for practice.

3.6   The difference in age between Brita and her daughter Eva is 30 years. The sum or their ages is 118. How old is Eva?

3.7   Jack is 3 years older than Susan. The sum of their ages is 27. Find their ages.

3.8   Glenn is 6 years older than Carla. The sum of twice Glenn's age and Carla's is 57 years. Find their ages.

3.9   Chris is 7 years younger than Mary. Chris's age subtracted from three times Mary's age is 43. Find their ages.

3.10 Ray is 2 years younger than Sig. Three times Sig's age subtracted from 5 times Ray's age equals Sig's age. How old are the boys?

Example:

The ratio of John's age to David's age is 6:5. In 7 years the ratio will be 7:6. What are their ages now?

A ratio is a comparison of two numbers by division. 6:5 is the same ratio as 12:10, 18:15, etc. Since the ratio is 6:5, ·we can call the ages $6x$ and $5x$. (See also Ratio problems in Chapter 5.)

Set up a table:

Name	Age now	Age in 7 years
John	$6x$	$6x + 7$
David	$5x$	$5x + 7$

The ratio of the ages is then 7:6 and we can set up an equation:

$$\frac{6x+7}{5x+7} = \frac{7}{6}$$

This is a proportion (two equal ratios), so we can cross multiply: $6(6x + 7) = 7(5x + 7)$
$$36x + 42 = 35x + 49$$
$$x = 7$$

John is $6(7) = 42$ years old and David is $5(7) = 35$ years old.

Check: In 7 years John will be 49 years old and David 42.
         Ratio: $49:42 = 7:6$

Alternate solution:

Name	Age now	Age in 7 years
John	x	$x + 7$
David	y	$y + 7$

Equations:
$$\frac{x}{y} = \frac{6}{5}$$
$$\frac{x+7}{y+7} = \frac{7}{6}$$

Cross multiply both equations:

$5x = 6y$ or $5x - 6y = 0$ $6(x + 7) = 7(y + 7)$

                                       $6x + 42 = 7y + 49$

                                       or $6x - 7y = 7$

We now have two equations:

$5x - 6y = 0$ Multiply by 7: $35x - 42y = 0$

$6x - 7y = 7$ Multiply by 6: $36x - 42y = 42$

Subtract the first equation from the second: $x = 42$

                                         $5(42) = 6y$

                                         $35 = y$

Practice Problems:

3.11 Adam and Victor are together 15 years old. Victor's age is 50% of his brother Adam's age. Find the ages of the boys.

3.12 The ages of the girls Ina, Mina, and Mo are in a ratio of 4:6:7. Combined the girls are 102 years old. What are their ages?

3.13 Mark and Mindy are together 84 years old. Three times Mark's age equals 4 times Mindy's age. How old are they?

3.14 The ages of Jim and Jon are in the ratio 3:7. If you multiply Jim's age by 4, you get the same result as when you add 40 to Jon's age. How old are Jim and Jon?

3.15 Six years ago, David's mother was 13 times as old as David. Now she is only 4 times as old as David. How old is David now?

Example:

Twice Lydia's age plus 1 equals 3 times her age less 4. How old is Lydia now?

If Lydia's age is x, then $\quad 2x + 1 = 3x - 4$

$$5 = x$$

Lydia is 5 years old.

Check: $\quad 2(5) + 1 = 10 + 1 = 11$

$$3(5) - 4 = 15 - 4 = 11$$

Practice Problems:

3.16 A mother is now 28 years older than her daughter. In 4 years the mother will be 3 times as old as the daughter. How old is the mother now?

3.17 Five years ago, Ellen's mother was 7 times as old as Ellen. Five years from now, she will be only 3 times as old as Ellen will be then. How old is Ellen now?

3.18 Ed is 8 years older than his brother John. Five years ago, Ed was 3 times as old as John. Find their present ages.

3.19 Carl is twice as old as Ginger. If Carl were 2 years younger and Ginger were 3 years older, the difference of their ages would be 3. How old is Ginger?

3.20 Ron is 6 years older than his wife, Bev. In 4 years, twice his age plus 1 will be 3 times Bev's age 3 years ago. How old are they now?

Example:

When Ralph asked his math teacher how old she was, she answered: "Ten years ago, my age was equal to the square of my daughter's age. In 14 years, I will be twice as old as my daughter."

Call the teacher's age x and her daughter's age y.

Make a table:

Name	Age now	Age 10 years ago	Age in 14 years
Teacher	x	$x - 10$	$x + 14$
Daughter	y	$y - 10$	$y + 14$

Equations: $x - 10 = (y - 10)^2 \rightarrow x - 10 = y^2 - 20y + 100$

$x + 14 = 2(y + 14) \rightarrow x + 14 = 2y + 28 \rightarrow x = 2y + 14$

Substitute x in the top equation with the expression for x in the second equation:

$$2y + 14 - 10 = y^2 - 20y + 100$$

Simplify: $y^2 - 20y + 100 = 2y + 14 - 10$

$$y^2 - 22y + 96 = 0$$

Factor: $(y - 6)(y - 16) = 0$

$$y - 6 = 0 \qquad y - 16 = 0$$

$$y = 6 \qquad\qquad y = 16$$

We must reject $y = 6$ because 10 years ago, the daughter would be –4.

We find the teacher's age by substituting in one of the original equations.

$x - 10 = (16 - 10)^2$

$x - 10 = 36$

$x = 46$

The teacher is 46 years old.

Check: In 14 years the teacher will be 60 years old and her daughter 30 years old.

Practice Problems:

3.21 The product of the ages of Dora and Phil is 243. Phil is 3 times as old as Dora. How old is Phil?

3.22 Liz is 2 years older than Ronald. The product of their ages is 440. How old are they?

3.23 Five years ago, the sum of the ages of Charlotte's daughters was 22. Five years from now, the older daughter will be exactly twice as old as the younger daughter. How old is the older daughter now?

3.24 A mathematician was asked how old she was. She answered: "I'll be x years old in the year x^2." How old will she be in the year 2006?
Hint: Use your calculator to find perfect squares in the years 2000 and later.

3.25 Susan is 5 years older than her sister Lucy. The sum of the square of Lucy's age and twice the age of Susan's is 58. How old is Lucy?

Chapter 4

Mixing Problems

In this chapter we learn to set up equations to solve word problems that deal with mixtures of various items. For example, stamps and coins of different denominations. We know the total number and value of a box full of nickels and dimes but want to find out how many nickels and how many dimes were mixed together in the box. Or we want to add water to a certain volume of concentrated fruit juice to make a larger volume of more dilute fruit juice. How much water should we add?

We will also deal with problems of mixing metals; for example, to calculate how much nickel a jeweler has to add to a quantity of pure gold (24-carat) to make 18-carat gold.

This chapter will also deal with situations involving differently priced products, such as adult and child admission tickets, apples and pears, gum and chocolates.

Finally, we will calculate what happens when one part of a sum of money is invested at one interest rate and the other part at a different rate.

Note: In all of these mixing problems, it helps to set up a table that lists all the given facts before trying a solution.

Stamps and Coins

Example:

Ed bought 150 stamps at the post office. Some were 37¢ stamps and some were 80¢ airmail stamps. He spent a total of $77. How many stamps of each kind did he buy?

Here we have a choice; one variable or two? Let's start with one.

Table:

Number of stamps	Value (¢)	Total Value (¢)
x	37	$37x$
$150 - x$	80	$80(150 - x)$
		7700

Equation:

$$37x + 80(150 - x) = 7700$$
$$37x + 12000 - 80x = 7700$$
$$4300 = 43x$$
$$100 = x$$

100 37¢ stamps

$150 - 100 = 50$ 80¢ stamps

If you prefer to work with two variables, we make a new table:

Number of stamps	Value (¢)	Total Value (¢)
x	37	$37x$
y	80	$80y$
150		7700

Equations: $x + y = 150$

$37x + 80y = 7700$

Multiply the first equation by 80 to get the same coefficient for y in both equations. Then, by subtracting one equation from the other, you eliminate the variable y:

$$80x + 80y = 12,000$$
$$37x + 80y = 7,700$$
$$\overline{43x = 4,300}$$
$$x = 100$$

Example:

Jessica sorted her grandma's coins. She found 96 coins, consisting of only nickels and pennies. If there was \$3.92 in total, how many nickels and how many pennies were there in the coin box?

Call the number of nickels x and the number of pennies $96 - x$. Then make a table:

	Number of coins	Value (¢)	Total value (¢)
Nickels	x	5	$5x$
Pennies	$96 - x$	1	$(96 - x)$
Total	96		392

Equation: $5x + 96 - x = 392$

$$4x = 296$$
$$x = 74$$
$$96 - 74 = 22$$

There were 74 nickels and 22 pennies.

Check: $74 \times 5 + 22 = 370 + 22 = 392$

Practice Problems:

4.1   Jens bought stamps at the post office. There were some $3.85 stamps for priority mail and some 80¢ stamps. He paid $162.00 for a total of 50 stamps. How many of each kind did he buy?

4.2   Michael has 60 coins consisting of nickels and dimes. Their total value was $5.00. How many of each coin does he have?

4.3   Sarah had $3.00 in nickels and dimes. There were 18 more nickels than dimes. How many coins are there of each kind?

4.4   A purse contains 27 coins, all quarters and dimes. The total value of the money is $6.45. How many coins are there of each kind?

Liquids With Different Strengths

When you mix two liquids, such as vinegar of different strengths, you get a vinegar solution with yet another strength. With these problems you set up a table.

Example:

Twenty ounces of vinegar with a strength of 30% was mixed with 40 ounces of a 20% vinegar solution. What was the percentage of the resulting solution?

Table:

Amount (oz.)	Strength (%)	Pure Vinegar (100%) (oz.)
20	30	20(30%)oz. = 6
40	20	40(20%)oz. = 8
60	x	6 oz. + 8 oz. = 14

Equation: $60(x\%) = 14$

$$60/100\, x = 14$$

$$x = 1400/60$$

$$x = 23.3$$

Answer: The new solution was 23.3%.

Practice Problems:

4.5 How many liters of a 20% alcohol solution must be added to 80 liters of a 50% alcohol solution to form a 45% solution?

4.6 How many ounces of pure acid (100%) should be added to 60 fluid ounces of a 20% acid solution to obtain a 40% acid solution?

4.7 Forty ounces of a punch containing 30% pomegranate juice was added to 50 ounces of a similar punch containing 10% juice. Find the percent pomegranate juice in the resulting punch.

Diluting Solutions With Water

In a pharmacy or in a chemistry lab one often dilutes a strong solution with water. You set up tables as above. Keep in mind that the strength of water is 0%.

Example:

A chemist needs a 10% alcohol solution and has only a 50% solution. How much water should be added to obtain 10 liters of the weaker solution?

Make a table:

	Amount (liters)	Percent (%)	Pure alcohol (100%)
Alcohol solution	x	50	$50\%x$
Water	$10-x$	0	0
	10	10	100%

Equation: $50\%x = 100\%$

$\qquad x = 2$

The chemist needs 2 liters of the 50% alcohol solution and 8 liters of water.

Check: 2 liters of 50% alcohol contains 2(50%)
$\qquad = 1$ liter pure alcohol.
1 liter pure alcohol diluted to 10 liters has a percentage $1/10 = 10\%$

Practice Problems:

4.8   How much water should be added to 4 liters of 90% alcohol to make a 40% alcohol solution?

4.9   How many liters of water should be added to 1 liter of a 20% saline (salt) solution to make a 5% saline solution?

4.10 How many quarts of 3% milk should be mixed with 100 quarts of 0.5% milk to make a 1.5% milk solution?

4.11 A pharmacist wants to dilute a 10% hydrogen peroxide solution to 3%. How much distilled water must he add to make 5 liters of 3% solution?

Mixing Metals

A jeweler often needs to mix pure metals (100%) such as gold or silver with other less costly metals. Here we also use tables very similar to the mixing of liquids.

Example:

How many grams of nickel should be added to 25 grams (g) of pure gold to give a metal that is 20% gold?

Make a table:

	Amount (g)	Percent (%)	Pure gold (g)
Nickel	x	0	0
Gold	25	100	25(100%) g = 25
Total	$x + 25$	20	$(x + 25)20\%$

Equation: $25 = (x + 25)20\%$

$$25 = 0.2x + 5$$
$$20 = 0.2x$$
$$x = 100$$

100 grams of the metal should be added.

Check: $(100 + 25)20\% = 125(0.2) = 25$

Practice Problems:

4.12 Silver is to be mixed with nickel. The jeweler specifies that the mixture should be 60% silver. How much nickel should be added to 90 g of silver?

4.13 Twenty-four-carat gold is the same as 100% pure gold. Suppose you have 300 g of 14-carat gold and want to increase the gold content to 18 carat, how much pure gold do you have to add?

A Mixed Bag

Many problems deal with the cost of tickets for shows where the adult price is more than that for children.

Example:

There were 250 people at a show in the local school. Adult tickets were $9 and tickets for children were $2.50. If $1112.50 worth of tickets were sold, how many adults and how many children attended the show?

Make a table:

	Number of people	Price of ticket	Total
Adults	x	$9	$9x$
Children	$250 - x$	$2.50	$(250 - x)2.50$
Total	250	—	$1112.50

Equation:
$$9x + (250 - x)2.50 = 1112.50$$
$$9x + 625 - 2.50x = 1112.50$$
$$6.5x = 487.5$$
$$x = 75$$
$$250 - 75 = 175$$

Answer: There were 75 adults and 175 children at the show.

Check: $75(9) + 175(2.50) = 675 + 437 = 1112.50$

The table looks more pleasant if we use both x and y:

	Number of people	Price of ticket	Total
Adults	x	$9	$9x$
Children	y	$2.50	$2.50y$
Total	250	—	$1112.50

Equations: $$x + y = 250$$ $$9x + 2.50y = 1112.5$$

Multiply the first equation by 2.5: $2.5x + 2.5y = 625$

Subtract: $$6.5x = 487.5$$ $$x = 75$$ $$y = 250 - 75$$ $$y = 175$$

We still get the same answer: There were 75 adults and 175 children at the show.

Practice Problems:

4.14 In an enclosure there are chickens and sheep. There is a total of 22 heads and 58 legs. How many chickens and how many sheep are there in the enclosure?

4.15 Selma has a lot of $10 and $20 bills. She has 5 times as many $10 bills as $20 bills. The sum of the money is $840. How many $10 bills does Selma have?

4.16 Ninety-five people bought a ticket to a concert. The tickets cost $10 each. The cashier counted 88 mixed $10 and $20 bills at the end of the concert. How many $20 bills were there?

4.17 The Carlson family bought charter tickets for 2 adults and 3 children. They paid $1450. An adult ticket cost $50 more than a child's. How much was an adult ticket?

Fruit, Candy, and Money

If we know the total price of a purchase and have enough information to get more than one equation, we can find the price of each item that was bought.

Example:

Two pounds (lb.) of pears and 3 lb. of apples together cost $4.26, whereas 3 lb. of pears and 2 lb. of apples cost $4.49. How much does each fruit cost per pound?

Because we have two unknowns, we need two equations and, therefore, two tables.

Item	Weight (lb.)	Cost/pound ($/lb.)	Total cost ($)
Pears	2	x	$2x$
Apples	3	y	$3y$
Total cost			$2x + 3y$ dollars or $4.26

Item	Weight (lb.)	Cost/pound ($/lb.)	Total cost ($)
Pears	3	x	$3x$
Apples	2	y	$2y$
Total cost			$3x + 2y$ dollars or $4.49

We have our two equations:

$2x + 3y = 4.26$ Multiply by 2. $4x + 6y = 8.52$

$3x + 2y = 4.49$ Multiply by 3. $9x + 6y = 13.47$

Subtract the first equation from the second: $5x = 4.95$

$x = 0.99$

The pears cost $0.99/lb.

Insert 0.99 in the first original equation:

$$2(0.99) + 3y = 4.26$$
$$3y = 4.26 - 1.98$$
$$3y = 2.28$$
$$y = 0.76 \qquad \text{The apples cost \$0.76/lb.}$$

Practice Problems:

4.18 Andy paid $1.70 for 5 pieces of gum and 8 small chocolates. Lisa bought 7 pieces of gum and 4 small chocolates. She paid $1.30. How much did the gum and the chocolates cost each?

4.19 Kelly works at a grocery store. In one week she earned $390 for 47 hours, of which 7 hours were overtime. The next week she earned $416 for 50 hours, of which 8 hours were overtime. What is Kelly's overtime rate?

Investments at Different Interest Rates

Example:

Ed invested $2000: one part at 3% interest and one part at 1.5% interest. If he got $41.25 simple interest in one year, how much did he invest at each percentage?

Make a table:

	Investment	Interest	Simple interest(%)
	x	3	3%x
	y	1.5	1.5%y
Total	$2000		41.25

Equations:	$x + y = 2000$
Solve for y:	$y = 2000 - x$
	$0.03x + 0.015y = 41.25$
Substitute:	$0.03x + 0.015(2000 - x) = 41.25$
Simplify:	$0.03x + 30 - 0.015x = 41.25$
	$0.015x = 11.25$
	$x = 750$
	So $2000 - 750 = 1250$

Answer: Ed invested $750 at 3% and $1250 at 1.5%.

Check: $750 \times 0.03 + 1250 \times 0.015 = 22.5 + 18.75$
$$= 41.25$$

Practice Problem:

4.20 Paul invested $3,000, part in a 1-year CD paying 1.20% and the rest in municipal bonds that pay 3% a year. The annual return from both accounts was $72. How much was invested in bonds?

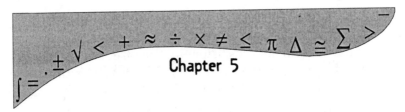

Chapter 5

Measurement Problems

In this chapter we will study different kinds of measurements. We will begin by comparing measurements in the form of ratios and proportions. For example, if you are twice as tall as your little sister, your height is to her height (in a ratio) as 2 is to 1. If your height is 6 feet (ft.) and your sister's is 3 ft., then 6 is to 3 as 2 is to 1 and that is a proportion. Other examples will show you how to determine the ratio of girls to boys or boys to girls in a class if you know how many of each there are, or to calculate the number of students and professors at a game if you know the total number and the ratio of students to professors.

Proportions are also useful to solve everyday problems: Say you earn $350 in a 5-day week, how long will you have to work to earn $2100? Or you know that your car gets 20 miles to the gallon, how many gallons of gas do you need to drive 1000 miles?

We will also use proportions to deal with indirect measurements. For example, to calculate the height of a flagpole by the length of its shadow if we know the height of a given object, say, a person, and the length of the object's shadow.

In this chapter, you will also learn the units used in measuring solids and liquids, areas and volumes, and also temperature. We will use the customary (English) system and also the metric system and show you how to convert from one to the other.

Ratio and Proportion

A *ratio* is a comparison of two numbers by division. The symbol for ratio is a colon (:) or a fraction bar (—) or (/).

Example:

Find the ratio of 2 feet to 6 feet.

$$2{:}6 \text{ or } \frac{2}{6} = \frac{1}{3} = 1/3$$

The ratio is 1 to 3.

Example:

Find the ratio of 2 ft. to 6 inches (in.).

When we compare two length measures, we usually convert the unit measures to the same kind.

2 ft. = 24 in.

$$\frac{24 \text{ in}}{6 \text{ in}} = \frac{4}{1} = 4/1$$

The ratio is 4 to 1.

Example:

In a class there are 15 boys and 18 girls. What is the ratio of *a*) boys to girls, *b*) girls to boys?

a) The ratio of 15 boys to 18 girls is $\dfrac{15}{18} = \dfrac{5}{6}$.

The ratio of boys to girls is 5:6.

b) The ratio of 18 girls to 15 boys is $\dfrac{18}{15} = \dfrac{6}{5}$.

The ratio of girls to boys is 6:5.

Example:

Two numbers are in the ratio 2 to 5. Their sum is 28. Find the numbers.

We call the numbers $2x$ and $5x$. No matter what x stands for, $2x$ and $5x$ always are in a ratio of 2 to 5 because we can reduce by x.

Equation: $2x + 5x = 28$

$$7x = 28 \text{ or } x = 4$$
$$2x = 2(4) = 8, \text{ and } 5x = 5(4) = 20$$

The numbers are 8 and 20.

Check: 8:20 (Reduce by 4) = 2:5, and $8 + 20 = 28$

Practice Problems:

5.1   Find the ratio of 5 minutes to 1 hour.

5.2   A 21-inch long ribbon is cut into 2 pieces in a ratio of 3 to 4. How long are the pieces?

5.3   In a town in Alaska, there are 3 women for every 7 men. How many men and how many women are there in the town if the total population is 4680?

5.4   The ratio of attendance at a college football game was 16 students for every professor. If a total of 3400 students and professors attended the game, how many of them were professors?

5.5   There are 8 women in a calculus class of 20 people.
Find the ratio of men to women in the class.

Proportion

Two equal ratios form a proportion. For example, $\dfrac{2}{3} = \dfrac{4}{6}$.
In a proportion, usually one of the numbers is unknown, so we
call that number x. If the example had been $\dfrac{2}{3} = \dfrac{x}{6}$, we would
have to solve for x.

In proportion problems we usually use cross multiplication,
so here we get $6(2) = 3(x)$. (If you need to review this, see the
Appendix for a review of equations.) This method can be used
only when we have a proportion, that is, two equal fractions.
We multiply one denominator by the opposite numerator, in
this case 6 times 2, then we multiply the other denominator
(here 3) by the opposite numerator (here x).

Solve the equation by dividing both sides by 3. The answer
is 4.

Example:

Pat earns $360 in 4 days. How many days will it take her to
earn $450?

Assume it will take x days.

Set up a proportion: $\dfrac{360}{4} = \dfrac{450}{x}$

Make sure that corresponding items occupy the same
places! Here we have both numerators showing the total
salary and both denominators the days. Case number
one is to the left and case number two to the right.

Cross multiply: $360x = 4(450)$

$x = 5$, so it takes Pat 5 days to earn $450.

If you prefer, you could set up the problem differently:

$$\frac{\$360}{\$450} = \frac{4}{x} \qquad \text{Reduce by 9:} \quad \frac{4}{5} = \frac{4}{x}$$

Just by looking at the equation, you can see that $x = 5$!

This problem could have been done in your head: $360 \div 4 = \$90$ and $\$450 \div \$90 = 5$. However, it is often useful to set up a problem formally for practice in order for you to solve more difficult problems you cannot do in your head.

Example:

Five quarts of ice cream were used to make 70 milkshakes. How many quarts of ice cream would be needed for 210 milkshakes?

Make a table:

	Case 1	Case 2
Quarts	5 quarts	x quarts
Shakes	70 shakes	210 shakes

Set up a proportion: $\dfrac{5}{70} = \dfrac{x}{210}$

Cross multiply: $5(210) = 70x$

$$x = 15$$

Answer: 15 quarts of ice cream are needed.

In your head you could think: 210 is 3 times 70, so we would need 3 times 5 quarts, which equals 15 quarts.

Practice Problems:

5.6   A sports utility vehicle can go 350 miles on 25 gallons of gasoline. How far can it go on 33 gallons of gasoline?

5.7   A 12-oz. can of frozen orange juice costs $2.56 and makes 48 oz. of diluted juice. What would the price be for a 6-oz. glass of diluted juice?

5.8   A photograph measuring 6 in. by 8 in. is to be enlarged so that the longer side is 12 in. What is the measure of the shorter side?

5.9   The ratio of two numbers is 4:5. Find the smaller number if the larger is 25.

5.10 A wheel turns 40 times every 2 minutes. At this speed, how many times will the wheel turn in 5 minutes?

Many proportion problems deal with indirect measurements. For instance, if you need to measure the height of a tall tree, instead of climbing up the tree with a huge tape measure, you can use the fact that sunrays are parallel. So if you can measure the length of the shadow and that of a known object, you can use similar right triangles for your investigation.

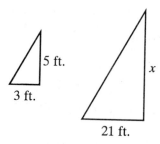

Figure 5.1

See also Chapter 8 (Geometry) about similar triangles.

Example:

A flagpole casts a shadow of 21 ft. while a boy 5 ft. tall casts a shadow of 3 ft. How tall is the flagpole?

Call the height of the flagpole x and set up a proportion:

$$\frac{x}{21} = \frac{5}{3}$$

Cross multiply: $3x = 5(21)$

$x = 35$, so the tree is 35 ft. tall.

Practice Problem:

5.11   Liam is 6 ft. tall. He wishes to know how tall the flagpole at his school is. He measures his own shadow, which is 8 ft., and the shadow of the flagpole, which is 48 ft. What is the height of the flagpole?

Measurements and Conversions

In the United States, daily life measurements are usually made in the *customary* system. Americans inherited the system from the British. Other countries, as well as most scientists, use the *metric*, or *S.I.* system, which the United States was supposed to adopt but has not yet.

The Customary System

The conversions among units in the customary system are as follows:

Length

1 mile = 5280 feet
1 yard = 3 feet
1 foot = 12 inches

Weight

1 pound = 16 ounces

Volume

1 gallon = 4 quarts
1 quart = 2 pints
1 pint = 2 cups
1 pint = 16 fluid ounces
1 cup = 8 fluid ounces

Example:

How many cups are there in 1 quart?

One quart equals 2 pints, which equals:
2×2 cups = 4 cups

There are 4 cups in one quart.

Practice Problems:

5.12   Convert 3 1/2 feet to inches.

5.13   Convert 12 ounces to pounds.

5.14   Convert 4 1/2 quarts to cups.

5.15   Convert 8 fluid ounces to pints.

Example:

Old people get shorter with age, so old Aunt Clara, who used to be 5 5/8 ft., shrank to 5 1/2 ft. How much did she shrink?

Here we have to subtract:

$$5\ 5/8\ \text{feet} - 5\ 1/2\ \text{feet} = \frac{45}{8} - \frac{11}{2} = \frac{45}{8} - \frac{44}{8} = \frac{1}{8}\ \text{ft. or}$$

$$\frac{12}{8} = \frac{3}{2} = 1.5\ \text{in.}$$

We could have solved this problem by first changing the feet to inches by multiplying by 12:

$$12 \times \frac{45}{8} - 12 \times \frac{11}{2} = \frac{3 \times 45}{2} - 6 \times 11 = 67.5 - 66 = 1.5$$

Aunt Clara shrunk 1.5 in.

Practice Problems:

5.16 You have a 3-yard long cloth. How much is left if you cut off 3 pieces that are 1.5 ft. long?

5.17 Michael is 6 7/12 years old, Jessica is 4 1/12 years old, and Kristina is 3 1/6 years old. What is the sum of their ages?

Example:

One room in your apartment is 12 ft. by 15 ft. The walls are 9.5 ft. high. How much paint will you need to paint the four walls, if one gallon is enough for 425 square feet (ft.²)?

Solution: You have four walls, two with a dimension of 12 by 9.5 ft.² and two of 15 by 9.5 ft.².
("By" signifies multiplication.)

The total area of the walls is:
$2 \times 12 \times 9.5 + 2 \times 15 \times 9.5 = 228 + 285 = 513$

If your algebra is strong, you could shorten the calculations to $2 \times 9.5(12 + 15) = 19 \times 27 = 513$.

One gallon of paint covers 425 ft.², x gallons cover 513 ft.².

Set up a proportion: $\dfrac{1}{425} = \dfrac{x}{513}$

Cross multiply: $513 = 425x$

$$x = 513/425$$

$$x = 1.2 \text{ (rounded)}$$

You need a little more than 1 gallon.

Practice Problems:

5.18 How many 1/2 oz. servings of butter can you get from a 2 1/2 lb. block of butter?

5.19 How many square yards of carpeting are needed for a room 24 ft. by 30 ft.?

5.20 Lyn walked 1 mile. How many yards is that?

5.21 To make a tart for 6 people, Karin needs 1 1/2 lb. of peaches and 2/3 cup of apricot jam. How much does she need for 20 people?

5.22 How many square inches are there in one square yard?

In the kitchen and also in hospitals other conversions are sometimes used. (See the chart on page 81.)

Example:

How many teaspoonfuls are there in 1 1/3 tablespoons?

One teaspoon is equivalent to 1/3 Tbsp. and 3 tsp. equals 1 Tbsp.; therefore, 4 tsp. equal 1 1/3 Tbsp.

Unit	Teaspoons	Tablespoons	Fluid Ounces
1 teaspoon	1	1/3	1/6
1 tablespoon	3	1	1/2
1 fluid ounce	6	2	1
1 cup	48	16	8
1 pint	NA	NA	16

Unit	Cups	Pints	Quarts
1 teaspoon	NA	NA	NA
1 tablespoon	1/16	1/32	NA
1 fluid ounce	1/8	1/16	1/32
1 cup	1	1/2	1/4
1 pint	2	1	1/2

NA = Not available. No values are listed in the original table.

Practice Problems:

5.23 Replace all NAs in the previous chart with numbers.

5.24 A recipe for 6 people calls for 5 Tbsp. of oil. How many cups are needed for 30 people?

5.25 A can of soda contains 12 fluid oz. How many cans must you buy to fill 10 8-oz. glasses?

5.26 A full bottle of cough syrup contains 8 fluid oz. You drink 2 Tbsp. How many Tbsp. of syrup are left?

5.27 A man donates 1 pint of blood to a blood bank. A blood transfusion may require 50 fluid oz. of blood. How many pints of blood are needed?

The Metric System

The *metric system* is built on powers of 10. For example, for length measures: 1 meter (m) = 10 decimeters (dm) = 100 centimeters (cm) = 1000 millimeters (mm) or written with powers: $1 \text{ m} = 10 \text{ dm} = 10^2 \text{ cm} = 10^3 \text{ mm}$.

Mass (weight) and volume are related in a similar fashion. The basic unit for length is the meter; for mass, the gram; and for volume, the liter. All other units are expressed with prefixes:

kilo (k) = thousand		10^3
hecto (h) = hundred		10^2
deka (da) = ten		10
deci (d) = tenth		10^{-1}
centi (c) = hundredth		10^{-2}
milli (m) = thousandth		10^{-3}

To convert between units we move the decimal point. Five kilometers (km) means 5×1000 meters = 5000 m. We move the decimal point 3 steps to the right: $5.000 \rightarrow 5000$.

Example:

2 dm = 20 cm = 200 mm

3 hg = 30 dag = 300 g

500 g = 5 hg = 0.5 kg

1000 mg = 100 cg = 10 dg = 1 g

Practice Problems:

5.28 Convert 4.5 kg to grams.

5.29 Convert 3.8 m to centimeters.

5.30 Convert 0.08 km to meters.

5.31 Convert 3890 g to kilograms.

5.32 Convert 150 mm to decimeters.

Example:

$1 \text{ dm} \times 1 \text{ dm} = 1 \text{ dm}^2$

$10 \text{ cm} \times 10 \text{ cm} = 100 \text{ cm}^2 = 1 \text{dm}^2$

Because all length measures are 10 apart, the area measures are 100 apart. For example, 100 square decimeters equal 1 square meter.

Practice Problems:

5.33 Convert 150 mm^2 to cm^2.

5.34 Convert 0.05 cm^2 to mm^2.

5.35 Convert 0.0075 m^2 to cm^2.

Example:

$1000 \text{ dm}^3 = 10 \text{ dm} \times 10 \text{ dm} \times 10 \text{ dm} = 1\text{m} \times 1\text{m} \times 1\text{m} = 1 \text{ m}^3$

The volume measures are 1000 apart.

For liquids, the measuring unit cubic decimeter has another name: liter (L).

$1 \text{ L} = 10 \text{ dL} = 100 \text{ cL} = 1000 \text{ mL}$

One liter and one quart of liquid are, more or less, the same.

Practice Problems:

5.36 Convert 3 m^3 to cubic decimeters.

5.37 Convert 5 dm^3 to cm^3.

5.38 Convert 0.5 L to cL.

5.39 A bathtub has a bottom area of 130 dm^2. How far up does the water rise if you put 390 L into the tub?

5.40 Butter, weighing 2 kg, is divided into small units of 1 hg each. How many small units will there be?

Example:

The scale of a map is 1/2 inch for every 3 miles. How far apart are 2 towns that are 4 1/4 inches apart on the map?

Table:

	Case 1	Case 2
Map	1/2 in.	4 1/4 in.
Real Life	3 miles	x miles

Set up a proportion:

$$\frac{\frac{1}{2}}{3} = \frac{4\frac{1}{4}}{x}$$

$$\frac{1}{2 \cdot 3} = \frac{\frac{17}{4}}{x}$$

$$\frac{1}{6} = \frac{17}{4x}$$

Cross multiply:

$$4x = 17 \cdot 6$$

$$4x = 102$$

$$x = 25.5 \text{ or } 25\,1/2$$

Answer: The distance is 25 1/2 miles.

Practice Problems:

5.41 A dining room is 15 m long and 12 m wide. Find the dimensions in a scale drawing which is 1:100.

5.42 On a drawing in a scale of 100:1 there is a distance, which is 125 mm. What is that in real life?

5.43 On a computer screen there is a 3-mm-long bug that has been enlarged in a scale of 200:1. What is the size of the bug on the screen?

Conversions Between the Customary and the Metric Systems

You need a table of conversion factors, but don't bother to memorize these factors. It is important to remember that 1 yard is a little less than 1 meter, 1 inch is about 2.5 centimeters, 1 pound is a little less than 0.5 kilogram, and 1 quart and 1 liter are more or less the same.

Length	Weight	Volume
1 in. = 2.54 cm	1 lb. = 454 g	1 qt. = 0.946 L
1 km = 0.6 mi	1 kg = 2.2 lb.	1 L = 1.06 qt
1 mi. = 1.6 km	1 oz. = 28 g	

Example:

Greg is 5 ft. 9 in. tall. What is his height in centimeters?

5 ft. 9 in. $= 5 \times 12 + 9 = 69$ in.

69×2.54 cm $= 175.26$

Answer: Greg is 175 cm. (The decimals are usually rounded.)

Example:

Convert 500 g to ounces.

$500/28 = 17.86 = 18$

Answer: 18 oz.

Practice Problems:

5.44 How many deciliters are there in 1 cup?

5.45 The new baby weighed 3.4 kg and was 51 cm long. Convert these measures to the customary system.

5.46 How many ounces are there in 100 g?

5.47 A mercury barometer reads 760 mm. Express this pressure reading in inches.

5.48 If a chemical costs $3.50 a pound, what is the cost of 15 g?

Dimensional Analysis

It is often confusing for people to convert from one unit to another, especially when many calculations are involved. In order to lessen the problems with conversions we use *dimensional analysis*. This is very important in scientific calculations.

Keep the measuring units with the numbers and treat them as if they are numbers that can be reduced. If, for example, you have *lb.* both in the numerator and the denominator, you simply cross them out. Remember that the "per" means division, so *$ per lb.* would be written as *$/lb.*

Example:

Convert $3.00 per pound to Swedish kronor per kilogram. There are 8.50 kronor in a dollar.

$$\frac{\$3}{\text{lb.}} \times \frac{8.50 \text{ kr}}{\$} \times \frac{2.2 \text{ lb.}}{\text{kg}}$$

Reduce: $$\frac{3 \times 8.50 \times 2.2 \text{ kr}}{\text{kg}} = \frac{56.1 \text{ kr}}{\text{kg}}$$

The cost is 56.1 kronor per kilogram.

Practice Problems:

5.49 An airplane flies with a speed of 900 km per hour. If you use a pen to follow its route on a map that has a scale of 1:50,000, what is the speed of the pen?

5.50 A skater once skated 10,000 m in a race in the time 14 min. 55.32 sec. Calculate the average speed in kilometers per hour. Round to a whole number.

5.51 If 150 g of cookies cost $1.25, how much do 5 oz. cost?

5.52 How many feet are there in a 500 m ice skating rink?

Temperature

In the customary system, temperature is measured in *Fahrenheit* (F) degrees and in the metric system, in *Celsius* (C) degrees. In the Fahrenheit system water boils at 212 degrees and freezes at 32 degrees. In the Celsius system water boils at 100 degrees and freezes at 0 degrees.

Conversion formulas:

$$\text{Formula 1: } C = \frac{5}{9}(F - 32)$$

$$\text{Formula 2: } F = \frac{9}{5}C + 32$$

Example:

Find the temperature in degrees Celsius if the temperature is 41 °F.

$$\text{Use Formula 1: } C = \frac{5}{9}(41 - 32) = \frac{5(9)}{9} = 5$$

The temperature is 5°C.

Approximation: Subtract 30 from 41 and divide by 2: $41 - 30 = 11$, and $11 \div 2 = 5.5$.

Example:

Find the temperature in degrees Fahrenheit when it is –5°C outside.

Use Formula 2: $F = \dfrac{9}{5}(-5) + 32 = -9 + 32 = 23$

Approximation: Multiply by 2 and add 30:
–5(2) + 30 = –10 + 30 = 20.

These shortcuts are very handy if you travel in a foreign country. If the temperature is either very high or very low, the approximation may not be accurate, but it is usually sufficient.

Practice Problems:

5.53 Calculate the temperature in Fahrenheit when the Celsius temperatures are: *a)* –40, *b)* 0, *c)* 10, and *d)* 100 degrees.

5.54 Calculate the temperature in Celsius when the Fahrenheit temperatures are: *a)* –40, *b)* –10, *c)* 32, *d)* 212 degrees.

5.55 A babysitter from Europe took the temperature of the little girl she was watching in the United States. It showed 100°. The babysitter got very frightened. Why was that?

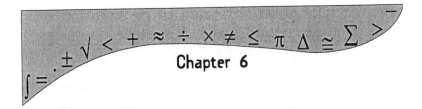

Chapter 6

Rate Problems

The word *rate* is frequently used in word problems. It is usually accompanied by another word, such as *per*, *for*, or *in*. For example: miles **per** hour, interest **per** year, price **per** pound or price **for** a certain weight, work done **per** hour or work done **in** so many hours.

In this chapter there will be examples of motion problems. When a car or person moves at a known speed (miles per hour) we can calculate the distance they will cover if we know the time it will take. Or knowing the distance and the time it takes, we can calculate the rate (speed) at which they travel. Or knowing the rate and time, we can calculate the distance.

We will also encounter problems where different objects (cars, boats, people) move at different rates and in different directions and where we want to calculate how far apart they will be in a given time or how long it will take them to cover a given distance. Or what happens when you row with the current of the river or against it.

In this chapter you will also learn how to solve work problems: how long it will take you to complete a certain task when you know how much you can do in an hour. For example, if you know that in one hour you can paint half the wall of your living room (that's your rate), how long will it take you to paint all four walls? Or if your friend agrees to help you and he paints twice as fast as you do, how long will it take?

Finally, there are the filling and emptying problems; such as, how long it takes to fill the bathtub when you forget to close the drain?

Motion (Speed) Problems

This category includes anything that moves; such as cars, boats, river currents, and airplanes.

Example:

Beth walked 2 miles in 40 minutes. What was her rate in miles per hour (mph)?

> Here we are given the time in minutes, so we have to convert that into hours.
>
> There are 60 minutes in 1 hour, so 40 minutes is 40/60 = 2/3 hour.
>
> To get speed per hour we divide 2 miles ÷ 2/3 hour:
>
> $$2 \div \frac{2}{3} = 2 \times \frac{3}{2} = 3$$
>
> Beth's rate was 3 mph.

Example:

Sheldon drove his car at an average of 55 mph. How far did he drive in 4 hours?

> The formula we can use is: **distance = rate × time**.
>
> $d = 55 \times 4 = 220$ miles

With help of dimensional analysis (see Chapter 5), we can check if our units are correct.

$$55\frac{\text{miles}}{\text{hour}} \times 4\,\text{hours} = 55 \times 4\,\text{miles}$$

(We reduce, or cancel, the unit *hours*.)

Example:

A car travels east at a speed of 50 mph and a motorcycle travels west at a speed of 60 mph. If they start at the same time and place, how long will it take them to be 220 miles apart?

Often with these types of problems it is useful to draw a picture.

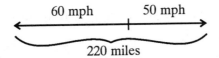

Figure 6.1

Make a table and recall that distance traveled equals time × rate:

Miles per hour is abbreviated as mph and

$$\frac{\text{miles}}{\text{hour}} \times \text{hours} = \text{miles}$$

	Rate (speed) (mph)	Time (hours)	Distance (miles)
Car	50	x	$50x$
Motorcycle	60	x	$60x$
Total			220

The *total distance* traveled is $50x + 60x = 220$.

Solve the equation: $110x = 220$ or $x = 2$

It took them 2 hours to be 220 miles apart.

Example:

Use dimensional analysis (Chapter 5) to convert 132 feet per 9 seconds to miles per hour.

Feet must be converted to miles and seconds to hours.

$$\frac{132\ \text{feet}}{9\ \text{seconds}} \times \frac{1\ \text{mile}}{5280\ \text{feet}} \times \frac{3600\ \text{seconds}}{\text{hour}} = \frac{132 \times 3600\ \text{miles}}{9 \times 5280\ \text{hours}}$$

$$= \frac{475200\ \text{miles}}{4750\ \text{hours}}$$

$$= \frac{10\ \text{miles}}{\text{hour}}$$

132 feet in 9 seconds equals 10 miles per hour.

Example:

Anne drove to her friend's house at a speed of 55 mph. During the trip back home, she could only drive 35 mph. It took her 2 hours longer to get home. How many miles is it between Anne's home and her friend's?

Make a table:

	Rate (mph)	Time (hours)	Distance (miles)
To the friend	55	x	$55x$
Back	35	$(x + 2)$	$35(x + 2)$

Note that the distance is the same going there as going home.

$55x = 35(x + 2)$

$55x = 35x + 70$

$20x = 70$

$x = 3.5$

Because it took her 3.5 hours to drive at a speed of 55 mph, the distance was:

3.5 × 55 miles = 192.5

Check: 35(3.5 + 2) miles = 35(5.5) miles = 192.5 miles

Practice Problems:

6.1 If you have walked 2.4 miles in 36 minutes, what was your average speed in miles per hour?

6.2 Carl and Bill live 54 miles from each other. One day they decided to bicycle from each others' homes. Carl bicycles with a speed of 15 mph and Bill with 12 mph. When will they meet and where?

6.3 Beatrice drives her car for 3 hours at a certain speed, and then increases the speed by 10 mph for the next 2 hours. The total distance she drove was 270 miles. Determine the two different speeds.

6.4 Hiking up a mountain took 5 hours, but down only 1 1/2 hours. If the distance each way was 3.3 miles, determine the difference between the two hiking rates.

6.5 Ben drove 200 miles at an average rate of 45 mph. On the return trip his average speed was 55 mph. What was the average speed for the total trip?

Example:

A bicyclist was training for a race. The training route was 75 miles. If the racer could increase his speed by 5 mph, he could complete the same course in 3/4 of the time he did earlier. Find his average rate of speed.

Table:

	Rate(mph)	Time(hours)	Distance(miles)
Case 1	x	y	$xy = 75$
Case 2	$(x + 5)$	$3/4\, y$	$(x + 5)\, 3/4\, y = 75$

We have two equations:

$xy = 75$

$(x + 5)\, 3/4\, y = 75 \rightarrow 3/4\, xy + 5(3/4)y = 75 \rightarrow$
$\quad 3xy + 15y = 300$ (Multiply both sides by 4.)

Substitute $xy = 75$ in the previous equation:

$3(75) + 15y = 300$

$\quad\quad 225 + y = 300$

$\quad\quad\quad 15y = 75$

$\quad\quad\quad\quad y = 5$

It took 5 hours to bicycle 75 miles, so his speed was 15 mph.

Practice Problems:

6.6   A car is driven first on a dirt road at a speed of 30 mph and then on a paved road with a speed of 60 mph. The total trip was 210 miles and took 4 hours. How long did the car travel on the dirt road?

6.7   Anna and Marie lived 36 km from each other. One day they decided to bicycle towards each other. Both started at 10:40 a.m., Anna with a speed of 12 km/h and Marie with a speed of 16 km/h. At what time did they meet?

6.8   The road between Cortland and Peekskill was old and crooked. After a new road was built, the distance between the towns was 10% shorter than before. The speed of the bus traveling between the towns increased by 20%. How much shorter (in percent) is the time it takes for the bus to travel between the two towns now?

Example:

It takes Jim 3 hours to row 12 miles with the current downstream and it takes him 6 hours to row the same distance against the current upstream. How fast can Jim row in still water? What is the speed of the current?

Think of what happens when he rows *with* the current: The current pushes him along so he will go faster. On the other hand, when he rows *against* the current, he is hindered by the current and goes slower. If Jim's speed is x mph and the current's y mph, then the speed with the current is $x + y$ and against the current $x - y$. (Don't write $y - x$, because then he would be going backwards!)

Make a table:

	Rate (mph)	Time (hours)	Distance (miles)
With	$x + y$ mph	3 hours	$(x + y)3$ miles $= 12$ miles
Against	$x - y$ mph	6 hours	$(x - y)6$ miles $= 12$ miles

Recall that parentheses followed or preceded by a number means multiplication.

We now have two unknowns and two equations:

$(x + y)3 = 12$ and $(x - y)6 = 12$

Solve this system of equations by dividing the first by 3 and the second by 6:

$$x + y = 4$$
$$x - y = 2$$
Add the equations:
$$2x = 6$$
$$x = 3$$
$$3 + y = 4$$
$$y = 1$$

Jim rows 3 mph in still water and the speed of the current is 1 mph.

Practice Problems:

6.9   Laura can row 16 miles downstream in 2 hours, but when she rows the same distance upstream it takes her 4 hours. Find Laura's rate in still water and the rate of the current.

6.10 A plane can travel 600 mph with the wind and 450 mph against the wind. Find the speed of the plane in still air and the speed of the wind.

Work Problems

Work problems are another type of *rate* problem. If we know how long it takes a worker to complete a certain task, we know the rate or which part of the job he can do per hour.

Example:

Lou can stamp 600 envelopes in two hours. What is his rate?

The total job takes 2 hours. Every hour he does 1/2 of the job. His rate is 1/2 of the total, or he can stamp $\frac{1}{2} \cdot 600 = 300$ envelopes per hour.

Example:

It takes Edwin 2 1/4 hours to paint a wall. What is his rate?

The total job takes 2 1/4 hours.

The rate per hour is: $\dfrac{1}{2\frac{1}{4}} = \dfrac{1}{\frac{9}{4}} = \dfrac{4}{9}$

He completes 4/9 of the job every hour.

Check: In 2 hours he completes 8/9 of the job and in 1/4 of an hour he completes the last 1/9 of the job.

Most work word problems deal with two people working together. Other problems deal with bathtubs that can be filled in a certain amount of time. Sometimes we leave the outlet plug open by mistake and then, how long does it take to fill the tub?

Example:

Anna can paint a room in 6 hours while Barbara can paint the same room in 4 hours. How long does it take if they work together?

Make a table:

Name	Rate/Hour	Time(hours)	Total Work
Anna	$\dfrac{1}{6}$	x	$\dfrac{x}{6}$
Barbara	$\dfrac{1}{4}$	x	$\dfrac{x}{4}$
Total			$\dfrac{x}{6} + \dfrac{x}{4} = 1$

The total job is 1.

$$\frac{x}{6} + \frac{x}{4} = 1 \quad LCD = 12 \text{ (The } least \ common \ denominator\text{)}.$$

Multiply both sides by 12: $2x + 3x = 12$

$$5x = 12$$

$$x = 2\frac{2}{5}$$

$$2\frac{2}{5} \text{ hours} = 2 \text{ hr.} + \frac{2}{5} \times 60 \text{ min.} = 2 \text{ hr. } 24 \text{ min.}$$

The total work time is 2 hours and 24 minutes.

Example:

Steven can wallpaper a room in 4 hours and Evie can wallpaper the same room in 6 hours. If Steven works alone for 1 hour and then both Steven and Evie work together to finish the room, how long will it take them assuming they keep their usual speed?

Make a table:

Name	Rate	Time	Total Work
Steven	1/4	$1 + x$	$(1 + x)1/4$
Evie	1/6	x	$1/6(x)$

Equation: $\dfrac{1+x}{4} + \dfrac{x}{6} = 1$ LCD = 12

$$(1 + x)3 + 2x = 12$$

$$3 + 3x + 2x = 12$$

$$5x = 9$$

$$x = 9/5$$

9/5 hours = 1 + 4/5 hour = 108 minutes

Practice Problems:

6.11 Repeat the previous example with Evie working 1 hour alone before Steven starts.

6.12 Jill can fill 300 envelopes per hour, whereas Fritz can fill this many envelopes in 2 hours. How long would it take for them to fill 300 envelopes if they work together?

6.13 It takes Nils 3 hours to mow the lawn but, with the help of Jonas, he can finish in 2 hours. How long would it take Jonas to do the job alone?

6.14 If Andrew, Bart, and Carl can wallpaper a house in 8 hours and Andrew and Bart can do it is 12, hours, how long will it take Carl to do the job alone?

6.15 Petra can do a certain job in 30 minutes and if she works together with Svea, they can do the same job in 24 minutes. How long would it take Svea to do it alone?

Example:

If 3 men can paint 4 houses in 5 days, how many days does it take for 7 men to paint 14 houses?

We have to suppose that all the men work at the same speed. Three men can paint 4/5 houses per day and one

man can paint $\dfrac{1}{3} \times \dfrac{4}{5} = \dfrac{4}{15}$ per day. At this rate it takes 7

men x days to paint 14 houses. That means that one

man can paint $\dfrac{1}{7} \times \dfrac{14}{x}$ per day.

These two expressions are equal: $\dfrac{4}{15} = \dfrac{1}{7} \times \dfrac{14}{x}$. Reduce

the right side of the equation and we get $\dfrac{4}{15} = \dfrac{1}{1} \times \dfrac{2}{x}$.

Cross multiply: $4x = 30$, so $x = 7.5$

It takes 7 1/2 days to paint these houses.

Practice Problem:

6.16 Robots are used to assemble automobiles in a factory. Three robots assemble 17 cars in 10 minutes. If all robots assemble cars at the same speed, how many cars can 14 robots assemble in 45 minutes?

Example:

It takes the hot water tap 30 minutes to fill the bathtub, while the cold water tap takes 20 minutes. How long would it take to fill the bathtub if both taps were open? Assume it takes x minutes.

Make a table:

	Rate	Time(minutes)	Part of bathtub filled
Hot water	1/30	x	$x/30$
Cold water	1/20	x	$x/20$

Equation: The bathtub is now filled, so we get the equation: $x/30 + x/20 = 1$

$$\frac{x}{30} + \frac{x}{20} = 1$$

Multiply both sides by 60 (LCD), simplify, and solve.

$$2x + 3x = 60$$
$$5x = 60$$
$$x = 12$$

Answer: It takes 12 minutes to fill the bathtub when both taps are open.

Example:

In the previous bathtub example, if the drain had been left open by mistake, how long will it take to fill the bathtub if the drain can empty the bathtub in 15 minutes?

As before, call the time x and make a table. Note that the rate when the bathtub is emptied is negative.

Make a table:

	Rate	Time (minutes)	Part of bathtub filled
Hot water	1/30	x	$x/30$
Cold water	1/20	x	$x/20$
Drain	−1/15	x	$-x/15$

Equation:

$$\frac{x}{30} + \frac{x}{20} - \frac{x}{15} = 1$$

Multiply both sides by 60 (LCD), simplify, and solve.

$$2x + 3x - 4x = 60$$
$$x = 60$$

Answer: With the drain open, it takes 60 minutes to fill the bathtub.

Practice Problems:

6.17 A tank can be filled by an inlet pipe in 15 minutes. A drain pipe can empty the tank in 60 minutes. How long does it take to fill the tank if both pipes are open?

6.18 A large tank can be filled by one pipe in 20 hours. A second pipe can fill the tank in 15 hours. If the first pipe was open for 16 hours and then closed, how long would it take to fill the tank if the second inlet pipe was opened after the first was closed?

6.19 It takes 12 minutes to fill a bathtub with water and half an hour to empty the bathtub through the outlet. How long will it take until the bathtub overflows if the inlet taps are open and the plug was not put in the outlet?

Chapter 7

Statistics and Probability

In this chapter we will solve problems that involve the organization and interpretation of numerical data. When numerical facts, such as the size of populations, the incomes of various groups, the number of cases of a disease, and the results on an exam, are collected, analyzed, and represented in the form of tables or graphs, we call it statistics.

We will begin with averages—for example, the average age of a group of students—and also calculate the median, mode, and range of this group.

Next, you will learn how to illustrate statistical data with bar graphs and pie diagrams, such as often seen in the newspapers or in a financial report from a company.

These bar graphs and pie diagrams can show you how test scores are distributed among different schools or what percentage of a club membership are men and women.

The next section of this chapter deals with probability and odds. When you draw a card out of a pack of cards, what is the probability that you will draw the queen of spades? Or when you roll dice, what are the chances that you will get two sixes? In this chapter you will find many problems that involve rolling a die, tossing pennies, and picking jellybeans out of a jar.

Finally, we get to permutations and combinations and sets. Take telephone numbers, which, when the area code is included, have 10 digits: In how many ways can these numbers be arranged? How many ways can the letters in a word such as MATHEMATICS be arranged?

Averages

Take, for example, the numbers 1, 2, 3, 4, 5. The *sum* of the numbers is 15 and there are 5 numbers. The *mean* of these numbers is $15 \div 5 = 3$. The *median* of the numbers is the middle number, or in this case is 3. The *mode* is the number that occurs most frequently. This set of numbers has no mode. The mean, median, and mode are all different kinds of averages. The most common average is the mean and that is what we use in daily life. When we ask, "What is the average age of this class?" what we want to do is find the mean. The *range* is the difference between the highest number and the lowest. It shows the spread of the data. In this case it is $5 - 1 = 4$.

Example:

The scores on a quiz are as follows: 6, 9, 8, 8, 7, 8, 10, 9, 7. Find the mean, median, and mode.

There are 9 entries, so the mean = $(6 + 9 + 8 + 8 + 7 + 8 + 10 + 9 + 7) \div 9 = 72 \div 9 = 8$.

To get the median we arrange the numbers in order: 6, 7, 7, 8, 8, 8, 9, 9, 10. There are 9 numbers, so the number in the middle is the fifth, or 8. The median is 8. If the number of entries is even, the median is considered to be between the two middle numbers, that is, the mean of these two numbers.

The number that occurs most frequently is 8, so the mode is also 8.

Mean, median, and mode are all 8.

The range is $10 - 6 = 4$.

Practice Problems:

7.1   Find the mean, median, mode(s), and range in the following sets of numbers:

  a) 9, 8, 2, 7, 1, 1, 5, 7

  b) 8, 9, 10, 8, 7, 7, 8, 6, 9, 10, 8, 7, 8, 9

  c) –2, –7, –5, 0, –5, –3

7.2   In a choir there were 12 men and 13 women. The mean average age of the men was 25 years and the mean average of the women 20 years. Find the mean age of the group. Hint: Find the total ages of the men and the total ages of the women first.

7.3   A 39-year-old man joins the group in practice problem 7.2. What is the new mean of the group?

7.4   In a group of 13 people the mean age is 25 years. A person who is 36 years leaves the group. What is the new mean of the group?

7.5   In a business with 45 employees the mean age is 36 years. Two 63-year-olds retire and a 24-year-old is hired. Find the new mean age.

Graphs

Statistical data are often illustrated with graphs. Take, for example, a bag of M&Ms that contains 50 candies: 20 light brown, 10 green, 5 red, 10 yellow, and 5 dark brown. This can be illustrated in a circle or pie graph the following way:

Light brown: $\dfrac{20}{50} = 40\%$

Green: $\dfrac{10}{50} = 20\%$

Red: $\dfrac{5}{50} = 10\%$

Yellow: $\dfrac{10}{50} = 20\%$

Dark brown: $\dfrac{5}{50} = 10\%$

To construct the circle graph, if 100% is the whole circle or 360°, then 40% is 144°, 20% is 72°, and 10% is 36°. These angles are called *central* angles, because they have their vertex at the center of the circle. To mark the angles on the circle you need a protractor.

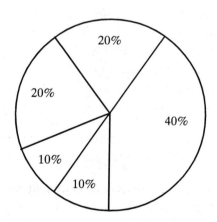

Figure 7.1

We will not have any exercises with making circle graphs here but will only learn how to read them.

The same data can also be displayed with a bar graph, where the horizontal line shows the different kinds of candies and the vertical line shows the percentages.

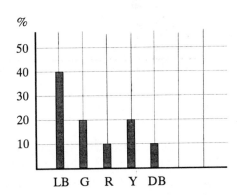

Figure 7.2

Example:

If we toss three coins, the following will happen: We get outcomes of 0, 1, 2, or 3 heads. Pretend that we toss a coin 16 times and get 0 heads twice, 1 head 5 times, 2 heads 8 times and 3 heads once. Prepare a table of the frequencies and make a bar graph and a frequency plot, also called a frequency polygon. The difference between a bar graph and a frequency polygon is that in the polygon we plot the frequencies as points.

Frequency Distribution

No. of heads	Frequency
0	2
1	5
2	8
3	1

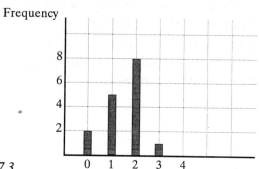

Figure 7.3

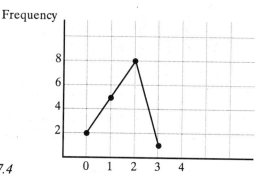

Figure 7.4

Practice Problems:

7.6 Toss 4 pennies 20 times and record the number of
heads you get. Make a bar graph and a frequency
polygon.

7.7 Find the mean, median, mode, and range for the
following data:

30	41	50	25	32
40	70	36	51	18
33	43	37	47	21

7.8 From the following frequency table of points on a test, find the mean, median, mode, and range.

Make a bar graph.

Number of points	Frequency
6	3
7	5
8	7
9	3
10	2

7.9 Twenty people were employed at a small factory. The number of sick days reported by each worker during a certain month were:

3, 3, 0, 1, 2, 5, 1, 2, 4, 3, 2, 2, 4, 2, 1, 6, 0, 2, 1, 1

Make a frequency distribution table and draw a bar graph. Find the mean and the median. What is the range?

7.10 The results of a 10-question quiz are as follows:

4	6	10	5
5	2	6	3
8	9	8	7
4	7	6	5
7	3	5	4

Make a bar graph and a frequency polygon of the scores. Find the mean, median, and mode.

Example:

From the following illustration, determine what percent of the members are girls?

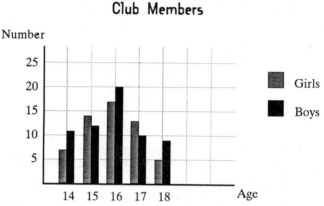

Figure 7.5

Read the bars and get the total number of each kind.

Girls: 7 + 14 + 17 + 13 + 5 = 56

Boys: 11 + 12 + 20 + 10 + 9 = 62

Total: 56 + 62 = 118

Girls are 56/118 = 0.47 = 47% (rounded).

There are 47% girls in the club.

Practice Problems:

7.11   The pie chart in Figure 7.6 shows the cost for the local school during one school year. There were 709 pupils that year and the total cost was $4.3 million. Determine the cost per pupil for *a*) salaries for teachers and *b*) upkeep of the buildings.

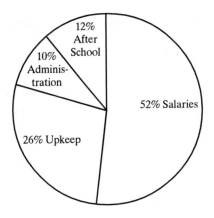

Figure 7.6

7.12 In a small country, the population of people over 65 years old was as follows: 652,735 men and 890,597 women. The total population that year was 8,837,496. Determine the central angle that shows this category of the population in a pie graph.

7.13 According to Figure 7.7, *a*) what was the amount of scholarship money awarded in 1986? *b*) by how much did the scholarship money increase between 1986 and 1989?

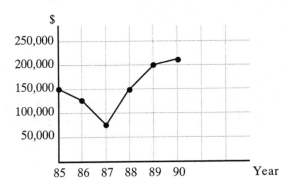

Figure 7.7

Probability and Odds

Probability

Probability is a measure of the likelihood of an event to happen. If something is absolutely (100%) sure to happen, the probability is 1 and if it is absolutely sure that it will not happen, the probability is 0. All other probabilities are expressed as fractions between 0 and 1. The probability of something to happen and not to happen adds up to 1.

The definition of probability is:

$$\frac{\text{number of favorable outcomes}}{\text{number of possible outcomes}}$$

Example:

A jar contains 60 jellybeans: 15 pink, 20 white, 5 yellow, and 20 purple. What is the probability of picking a purple jellybean?

> There are 20 purple jellybeans out of a total of 60; therefore, the probability of picking a purple jellybean is 20/60 = 1/3.
>
> The probability is 1/3.

Practice Problems:

7.14   What is the probability of picking *a*) a pink, *b*) a white, *c*) a yellow?

7.15   There are 6 white and 4 black balls in a box. What is the probability of picking up a white ball without looking?

7.16   A digital clock shows the time every minute. What is the probability that during a 24-hour period when you look at the clock all digits showing the time are the same?

Example:

A deck of cards consists of 52 cards. You draw one card. What is the probability that the card is a king?

To solve this problem you need to know that there are 4 kings in a deck. The probability is 4/52 = 1/13.

Example:

Find the probability of drawing a red card from a deck.

There are 26 red and 26 black cards in a deck. Therefore, the probability of drawing a red card is 26/52 or 1/2.

Example:

Find the probability of drawing a face card.

Face cards are the cards with a picture of a king, queen, or a jack rather than a card with pictures of clubs, spades, etc., and numbers. The ace stands for either 1 or 14.

There are $3 \times 4 = 12$ face cards in a deck; 3 from each suit and 4 suits.

12/52 = 3/13

The probability of getting a face card is 3/13.

Practice Problems:

7.17 Find the probability of drawing a queen of spades from a deck of cards.

7.18 Find the probability of drawing a red face card from a deck of cards.

7.19 A die has 6 faces representing the quantities 1 through 6. If you roll a die, what is the probability of getting *a*) a 4, *b*) an odd number, *c*) a number less than 3, *d*) a number more than 6?

Example:

If a penny is tossed twice, what is the probability of two heads?

The possible outcomes are: head head, head tail, tail head, tail tail.

Two heads occur once out of four.

The probability of getting two heads is 1/4.

Practice Problems:

7.20 If a penny is tossed three times, what is the probability of *a*) three heads, *b*) two heads, *c*) no heads?

7.21 A piggy bank contains 40 pennies, 25 nickels, 15 dimes, and 4 quarters. Find the probability that the coin is: *a*) a nickel, *b*) a penny, *c*) a nickel or a quarter.

Odds

Odds can be in favor of an event or against it.
The definition of odds in favor is:

$$\frac{\text{number of favorable outcomes}}{\text{number of unfavorable outcomes}}$$

Example:

What are the odds in favor of getting a 3 in one roll of a die?

There is 1 favorable way of getting a 3 and 5 unfavorable ways. Consequently, the odds are 1 to 5.

What are the odds against getting a 3?

There are 5 unfavorable ways and 1 favorable. The odds are 5 to 1.

Practice Problems:

7.22 On one roll of a die, what are the odds in favor of getting an even number?

7.23 On one roll of a die, what are the odds against getting an odd number?

Probabilities With "And" and "Or" Statements

If you toss a penny, you get either a head (H) or a tail (T). That is called the outcome. What is the probability of getting a head on your first toss? You have two possibilities: head or tail. One event (H) is favorable to you. The probability is 1/2. If you toss one penny twice or two pennies once, you get the following outcomes: HH, HT, TH, TT. What is the probability of getting 2 heads? One outcome (HH) out of 4 is favorable. The probability is 1/4.

The problem could be stated: What is the probability of getting first a head **and** then a second head? $1/2 \times 1/2 = 1/4$. The word "and" translates into multiplication.

Example:

You draw a card from a standard deck of cards and then put it back. Now draw another card from the same deck. What is the probability that you get a face card both times?

There are 3 face cards of each suit in a deck of cards or 12 in the whole deck, which consists of 52 cards. The probability of picking one face card is 12/52, or 3/13. The probability of picking first a face card and then another face card is $3/13 \times 3/13 = 9/169$.

Note: If the first card was not put back into the deck, the second probability would have been 11/51, because we removed one face card from the deck.

Practice Problems:

7.24 Find the probability to get first a 6 and then a 3 when you toss a die.

7.25 The probability that a certain event occurs is 30%. What is the probability that the event occurs twice in a row? Give the answer in percent.

7.26 Find the probability that you get first get a 1 and then and even number when you roll a die.

7.27 You toss a coin and roll a die. What is the probability that you get a head and a 5?

7.28 A jar contains 3 red marbles, 2 white, and 4 blue. A marble is chosen at random from the jar and then replaced. Another marble is chosen. *a*) Find the probability that both marbles are white. *b*) If the first marble is not replaced, find the probability that both marbles are blue.

Suppose the problem was: What is the probability of getting a head on the first toss **or** on the second toss (I don't care which one) when I toss a penny? The possible outcomes of two tosses are HH, HT, TH, and TT. The probability of each outcome is 1/4. One head occurs two times out of four, so the probability is 2/4 = 1/2.

The word "or" indicates that we should add.

Example:

What is the probability of getting 1 or 2 heads when tossing two pennies?

The probability of 2 heads is written as p(HH) and equals 1/4. p(HT) = 1/4 and p(TH) = 1/4.
1/4 + 1/4 + 1/4 = 3/4

The probability of getting one or two heads is 3/4.

Practice Problems:

7.29 There are 8 green, 7 red, and 10 blue marbles in a box. What is the probability that the marble is red or blue if you pick up a marble without looking?

7.30 What is the probability to get a king or a queen when you pick a card from a deck?

7.31 What is the probability of obtaining a 1 or an even number on a single roll of a die?

Example:

What is the probability of obtaining a 2 or an even number on a single roll of a die? Here we have a complication. Because 2 is an even number it is counted twice. We have to subtract the probability 1/6 from the sum of the probability of rolling a 2, which is 1/6, and the probability of getting an even number, which is 3/6.

The answer is $1/6 + 3/6 - 1/6 = 3/6 = 1/2$.

Practice Problems:

7.32 If two dice are rolled, what is the probability that the outcome is either doubles or that the sum will be 10?

7.33 Find the probability that when a die is tossed, you get an odd number or nothing lower than a 5.

7.34 When you roll a die, what is the probability to first get a 3 and then a 1 or a 2?

The Counting Principle, Permutations, and Combinations

The Fundamental Counting Principle

Assume that there are three ways for an event to happen and two ways for another event to happen, then the fundamental counting principle tells us that there are $3 \times 2 = 6$ ways for both events to happen. If you have 3 pairs of pants and 8 shirts, how many outfits can you arrange? Each pair of pants can be combined with 8 different shirts, so 3 pairs of pants can be combined with 8 shirts to form $3 \times 8 = 24$ different outfits.

Example:

Suppose you are taking a multiple choice test with 5 questions and there are 4 choices for each answer. How many different test results can you have?

The first question has 4 possible answers.

The second question has 4 possible answers.

The third question has 4 possible answers.

The fourth question has 4 possible answers.

The fifth question has 4 possible answers.

In total there are $4 \times 4 \times 4 \times 4 \times 4 = 1024$ possible test results.

Practice Problems:

7.35 You have a dating service with 534 women and 216 men. How many different dates can you arrange?

7.36 How many three-digit numbers can be formed from the digits 2, 3, and 4 if repetition of each digit is allowed?

Permutations

A permutation is an arrangement of things in a definite order. For example, if we have to arrange 4 people in a row, we can put any one of the 4 people in the first position, one of the 3 remaining people in the next, 2 after that, and 1 in the last position. The fundamental counting principle tells us to multiply, so we get $4 \times 3 \times 2 \times 1 = 24$ different ways of arranging these 4 people.

When we work with permutations and combinations we often use a mathematical concept called a factorial. It is written as $n!$ and pronounced n-factorial. When $n = 4$:

$$4! = 4 \times 3 \times 2 \times 1 = 24$$

If we have to arrange n things but a of them are alike, we must divide $n!$ by $a!$

Example:

In how many ways can 3, 5, 6, 6, 7 be arranged?

There are 5 numbers with 2 of them that are alike:

$$5!/2! = 5 \times 4 \times 3 \times 2 \times 1 / (2 \times 1) = 60$$

Practice Problems:

7.37 In how many ways can the names of 5 candidates for the same office be listed on a ballot?

7.38 In how many ways can the digits 2, 3, 5, and 7 be arranged so that the number is less than 4000 when the digits cannot be repeated?

7.39 How many even 4-digit numbers can be formed using the digits 1, 2, 3, and 9 if repetition of digits is not allowed? *Hint*: Determine which digit must be the fourth.

7.40 The letters in the word CIRCLE are randomly rearranged. In how many different ways can the letters be arranged?

Combinations

A combination is a selection of things in which order does not matter. If the letters A, B, and C are selected two at a time, we get three possibilities: AB, AC, and BC. We can use a formula $n!/r!(n-r)!$, where n is the number of objects and r the number of objects selected at a time. In our case $n = 3$ and $r = 2$.

$$3!/2! = 3 \times 2 \times 1 /(2 \times 1) = 3$$

We get 3 combinations of the letters A, B, and C.

Sets

A *set* is a collection of things called *elements*. They can be listed the following way: A set containing the numbers 1, 2, and 3 is put in *braces*: $\{1, 2, 3\}$. If a set contains an infinite number of elements such as the counting numbers, which are the numbers we count with, we use dots to signify that there are more numbers than we can list: $\{1, 2, 3, ...\}$.

Practice Problems:

7.41 Write in set notation the set of all positive even integers.

Two sets can *intersect*, meaning that there are elements that the sets have in common. Symbol: $\cap$.

If two sets do not have any elements in common, the result is written $\{ \}$ or $\varnothing$. It is called the empty set.

Example:

Find the intersection of the sets $\{1, 2, 3, 4, 5\}$ and $\{4, 5, 6\}$.

$$\{1, 2, 3, 4, 5\} \cap \{4, 5, 6\} = \{4, 5\}$$

The intersection is $\{4, 5\}$.

When we list all elements of two sets without repeating common elements, we get the *union* of the sets. Symbol: ∪.

Example:

Find {1, 2, 3, 4, 5} ∪ {4, 5, 6}.

The union is {1, 2, 3, 4, 5, 6}.

The intersection and union of two sets are often illustrated by *Venn diagrams*. These are circles representing the sets. See Figure 7.8.

Example:

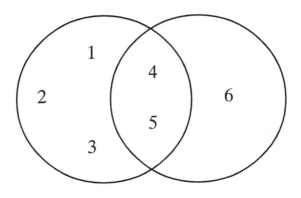

{1, 2, 3, 4, 5} ∩ {4, 5, 6}

Figure 7.8

Practice Problems:

7.42 Find *a*) the intersection and *b*) the union of the sets
$A = \{1, 3, 5\}$ and $B = \{2, 4, 6, 8\}$.

7.43 In a school there are 387 juniors: 165 take both chemistry and German, and 258 take Biology and German. How many students are taking German?

7.44 At a musical audition of 91 people, 32 people said that they could sing but not dance, 45 could dance but not sing. How many people could both sing and dance?

7.45 A car salesman sells 150 cars in a certain month. Out of these cars 90 have an airbag on the passenger side, 50 have a car phone, and 30 have neither an airbag nor a car phone. How many cars have both an airbag and a car phone?

Chapter 8

Geometry

In this chapter you will find an introduction to the basics of plane and solid geometry, trigonometry, and analytic geometry; that is, the world of angles and triangles, rectangles, squares, circles, cubes, and cylinders. We will not have to become involved with proofs, but only with word problems that ask for numerical answers.

In *plane geometry*, we will calculate areas of triangles and other planar figures and their perimeters, circumferences, diameters, and radii of circles, and, later on, the volumes of cubes and cylinders.

You will learn about the Pythagorean Theorem and how to apply it. You will become acquainted with similar triangles and how they are used for various calculations.

Then we will move onto problems involving figures with more than three sides, such as quadrilaterals, polygons (for example, the Pentagon in Washington, D.C., with five angles and five sides, is a polygon), and also circles.

In *solid geometry* you will calculate areas and volumes of different solid figures, such as cubes and cylinders. In the section on *trigonometry* there will be problems involving the three important functions sine, cosine, and tangent and how to find their values using a calculator. Finally, in the section on *analytic geometry* (also called coordinate geometry), you will learn how to represent points and equations by their coordinates in a graph and how to calculate distances between points or areas of triangles and rectangles formed by these points.

Plane Geometry

This chapter deals with the basics of geometry. There are no proofs, only calculations. In plane geometry we deal with *lines*, *rays*, and *line segments*.

A straight line has no beginning, no end, and no size, only direction. It is represented by a line with arrows. A ray is a part of a line with a beginning but no end. A line segment is a part of a line with a beginning and an end.

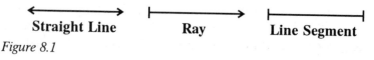

Straight Line   **Ray**   **Line Segment**

Figure 8.1

Angles

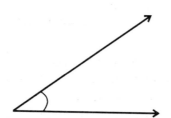

Figure 8.2

Angles are formed by two rays with the same endpoint (*vertex*). They are measured in *degrees*. If the rays coincide, the angle measure is 0°. When the rays form a straight line, the angle measure is 180°: This is a *straight* angle. When the angles are *perpendicular*, the angle measure is 90°: This is a *right* angle. A right angle looks like the corner of a book or any other rectangular object. Symbol for angle: $\angle$.

An angle between 0° and 90° is *acute* and an angle between 90° and 180° is *obtuse*. When two angles' measurements add up to 180°, they are *supplementary*, and two angles whose measurements add up to 90° are *complementary*.

Example:

Find the supplementary angle to 135°.

Two supplementary angles add up to 180°, so $180 - 135 = 45$.

The angle is 45°.

Practice Problems:

8.1 Find the complementary and supplementary angles to *a*) 45°, *b*) 30°, *c*) 90°, *d*) 67°, and *e*) 75°.

Perimeter

The perimeter of a plane figure is the total length around it. Think of measuring it with a tape measure.

Example:

A rectangle has a perimeter of 72 inches. If the length is 6 inches more than the width, find the length and width.

Call the width x. Then the length is $x + 6$ and the perimeter is

$$x + x + 6 + x + x + 6 = 72$$
$$4x + 12 = 72$$
$$4x = 60$$
$$x = 15$$
$$x + 6 = 21$$

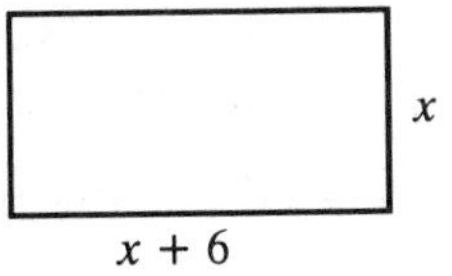

Figure 8.3

The width is 15 inches and the length is 21 inches.

Check: $15 + 21 + 15 + 21 = 72$

Practice Problems:

8.2   Find the dimension of a rectangle whose length is twice the width and whose perimeter is 42 cm.

8.3   The length of a rectangle is 6 units more than its width. If the perimeter is 40 units, find the length and the width.

Example:

The base of an isosceles triangle is 7 inches and the perimeter is 29 inches. Find the other sides.

An *isosceles triangle* has two equal sides. The third side is called the *base*.

Call each of the equal sides x.

The perimeter is $x + x + 7 = 29$.

$$2x + 7 = 29$$
$$2x = 22$$
$$x = 11$$

The sides are each 11 inches.

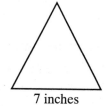
7 inches

Figure 8.4

Practice Problems:

8.4   A rectangle has a perimeter of 62 cm. The length of the rectangle is 5 cm longer than the width. Find the length and the width of the rectangle.

8.5   Find the perimeter of a square with a side of 5 inches.

8.6   The width in a rectangle is 2/3 of the length and the perimeter is 15 inches. Find the length and width of the rectangle.

8.7 An equilateral triangle (all sides are equal) has the same perimeter as a square. The side of the triangle is 1.2 inches longer than the side of the square. Find the perimeter.

8.8 In the triangle *ABC*, the side *AC* is 8 cm longer than side *AB*. The side *BC* is twice the side *AB*. The perimeter of the triangle is 60 cm. How long are the sides?

Areas

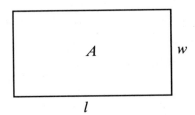

Figure 8.5

The *area* of a *rectangle* is length × width.

Example:

A rectangle has a length of 10 cm and a width of 6 cm. Find the area.

$l \times w = 10$ cm × 6 cm = 60 cm² (square centimeters). Note that all areas are expressed in square units.

Example:

A square has a perimeter of 24 inches. Find the area.

A square is a rectangle with all sides equal. The side is usually called *s*, so the area is $s \times s$, or s^2.

$4s = 24$

$s = 6$

$A = 6 \times 6 = 36$ in.²

The area is 36 square inches.

Practice Problems:

8.9   A square has an area of 81 in.². What is the perimeter?

8.10 A square has a perimeter that is 3 inches longer than the perimeter of an equilateral triangle. The side of the triangle is 1/2 inch longer than that of the square. Determine the area of the square.

8.11 Two of the sides of a square are increased by 2 inches and the other two sides are shortened by 1 inch. The new rectangle has an area that is equal to the original square. Find the area.

8.12 The perimeter of a certain rectangle is 24 inches. If the length is doubled and the width is tripled, the area is increased by 160 in.². Find the dimension of the original rectangle.

8.13 A square has an area of 64 in.². Two of the sides of the square are increased by 40% and the other two sides are shortened by 25%. A rectangle is formed. How much larger (in percent) is the rectangle than the square?

The *area of a triangle* is $A = 1/2bh$ where b is the *base* and h is the *height* (or *altitude*), which is a line segment from the *vertex* (corner) opposite the base and perpendicular (under right angles) to the base. See Figure 8.6.

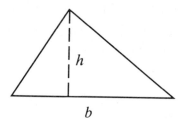

Figure 8.6 b

Any side of the triangle can be considered the base. Each base has an altitude connected to it.

Example:

Find the area of a triangle with a base of 4 inches and the height of 3 inches.

Formula: $A = \dfrac{1}{2}bh$

$A = \dfrac{1}{2}(4)(3)$ in.^{2} = 6 in.^{2}

The area is 6 square inches.

Example:

The legs (the sides that form the right angle) of a right triangle are 5 cm and 8 cm. Find the area.

If one leg of the right triangle is taken as the base, the other leg is the height.

Area: $\dfrac{1}{2} \cdot 5 \cdot 8 = 20$

The area is 20 cm^2.

Practice Problems:

8.14 Two sides of a triangle are 15 inches and 20 inches. The altitude to the longer side is 6 inches. Find the altitude to the shorter side.

8.15 A triangle has an area of 30 in.^{2} and an altitude of 5 inches. Find the base belonging to this altitude.

The Pythagorean Theorem

If the legs of a right triangle are a and b, we have the relationship $a^2 + b^2 = c^2$, where c is the *hypotenuse* (the longest side). See Figure 8.7.

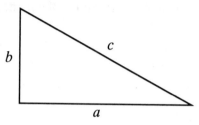

Figure 8.7

Example:

Find the hypotenuse if the legs are 3 inches and 4 inches.

$3^2 + 4^2 = 9 + 16 = 25 = 5^2$

Answer: The hypotenuse is 5 inches.

Example:

Find the other leg, if one leg is 5 cm and the hypotenuse is 13 cm.

Call the missing side x.

$$x^2 + 5^2 = 13^2$$
$$x^2 + 25 = 169$$
$$x^2 = 144$$
$$x = 12$$

The leg is 12 cm.

Practice Problems:

8.16 Find the hypotenuse if the legs are 6 and 8 units each.

8.17 Find the hypotenuse if the legs are 7 and 24 units each.

8.18 Find the area of a triangle with sides 3 in., 4 in., and 5 in. The altitude to the longest side is 2.4 in. Can you find two ways of solving this problem? *Hint:* What kind of triangle is this?

8.19 Find the missing leg if one leg is 7 and the hypotenuse is $\sqrt{65}$ units.

8.20 A 20-ft.-long ladder stands against a wall. The bottom of the ladder is 8 ft. from the wall. How high up on the wall is the ladder?

Angles and Triangles

The sum of the angles in any triangle is 180°.

Example:

If the angles in a triangle are x, $2x$, and $3x$, find x.

$$x + 2x + 3x = 180$$
$$6x = 180$$
$$x = 30$$

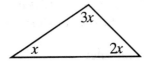

Figure 8.8

The angle marked x is 30°. The other angles are 60° and 90°.

Practice Problems:

8.21 In a triangle the angle A is twice angle B. Angle C is 20° more than angle B. Find the angles.

8.22 In the triangle ABC, the angle A is three times angle B. Angle C is 60°. Find angle A.

8.23 The ratio of two angles in a triangle is 3:5. The third angle is 52°. Find the other two angles.

Exterior Angles

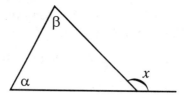

Figure 8.9

If one side of a triangle is extended, the angle formed between the extended side and the other side is called *exterior*.

An exterior angle is equal to the *sum of the two non-adjacent angles*. In other words,

$$x = \alpha + \beta$$

Example:

How many exterior angles does a triangle have?

Each vertex (corner) can create two exterior angles, and they are always equal; therefore, there are at most three different exterior angles.

Example:

Find the exterior angles in a triangle where the angles are 37°, 42°, and 101°.

$37° + 42° = 79°$

$37° + 101° = 138°$

$42° + 101° = 143°$

The exterior angles are 79°, 138°, and 143°.

What is the sum of the three exterior angles in the previous example? Is that true for all triangles?

Example:

Find the angles of the following triangle: One exterior angle is $14x - 6°$ and the two non-adjacent interior angles are $7x$ and $5x + 10°$.

$$14x - 6 = 7x + 5x + 10$$
$$14x - 6 = 12x + 10$$
$$2x = 16$$
$$x = 8$$

Exterior: $14(8°) - 6° = 106°$

Non-adjacent interior: $7(8°) = 56°$

Non-adjacent interior: $5(8°) + 10° = 40° + 10° = 50°$

Third interior: $180° - 106° = 74°$

The angles are $74°$, $56°$, and $50°$.

Practice Problems:

8.24 Triangle ABC is isosceles where A and C are base angles. The exterior angle at C is $156°$. Find angle B.

8.25 In the triangle ABC, side $\overline{AB}$ is extended to D. If the angle at C is $x + 30$, the angle at A is $2x + 10$, and the exterior angle at B (CBD) is $4x + 30$, what is the value of x? Note: When you write $\overline{AB}$ you imply that you deal with a line segment. In other words, it is the physical side between A and B. AB without the bar means the measure of the line segment.

Congruent and Similar Triangles

If two triangles are equal in both shape and size, they are *congruent*. If one triangle were cut out and moved, it would fit exactly over the other. Symbol: $\cong$.

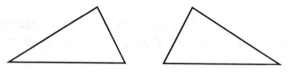

Figure 8.10

The concept of congruence is used mainly in proofs. The concept of similarity is more useful for word problems. *Similar figures* have the same shape but not the same size. Symbol: ~.

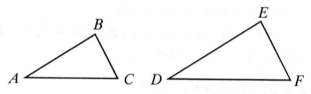

Figure 8.11

Similar triangles have the corresponding angles equal and corresponding sides proportional.

For example, if triangle *ABC* ~ triangle *DEF*, then

$$\angle A \cong \angle D \quad \angle B \cong \angle E \quad \angle C \cong \angle F$$

$$\frac{AB}{DE} = \frac{BC}{EF} = \frac{AC}{DF}$$

Example:

In the previous example, if $AB = 3, BC = 4, AC = 5, DE = 6$, what are *EF* and *DF*?

$$\frac{AB}{DE} = \frac{3}{6} = \frac{1}{2}$$

$$\frac{BC}{EF} = \frac{4}{x} = \frac{1}{2}$$

$$x = 8$$

$$\frac{AC}{DF} = \frac{5}{y} = \frac{1}{2}$$

$$y = 10$$

$EF = 8$ and $DF = 10$

Practice Problems:

8.26 In triangle ABC, $AB = 3$, and $BC = 2$ and in the similar triangle DEF the corresponding sides are x and 3. Find x.

8.27 The sides of a triangle are 7 cm, 10 cm, and 12 cm. In another triangle, similar to the first, the shortest side is 10.5 cm. Find the other sides in the second triangle.

8.28 A boy if 6 ft. tall and his shadow measures 4 ft. At the same time of day, a tree's shadow is 24 ft. long. How tall is the tree? *Hint*: Use two similar right triangles.

Polygons

The word polygon means many angles. But it is also used to mean many sides. The simplest polygons are triangles and then quadrilaterals (four-sided). Here belong the square, the rectangle, the rhombus, the parallelogram, and the trapezoid.

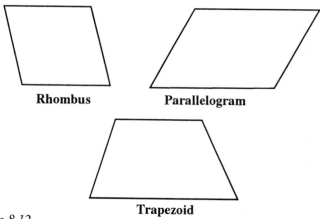

Rhombus   **Parallelogram**

Trapezoid

Figure 8.12

Similar Polygons

If a polygon is enlarged, all sides are enlarged in the same proportion. What happens to the area?

As an example, take a 2×3 rectangle. Its area is 2 times $3 = 6$ square units.

Multiply each side by 2. The new rectangle has a width of 4 units and a length of 6 units. Its area is 24 square units, that is, 4 times the original rectangle.

Example:

The ratio of the perimeters in two polygons is 2:3. If the area of the small polygon is 5 in.², find the area of the large polygon.

The ratio of the areas is the square of 2:3 or 4:9.

$$\frac{4}{9} = \frac{5}{x}$$

Cross multiply: $4x = 45$

$$x = 11.25$$

The area is 11.25 in.².

Practice Problem:

8.29 A triangle has an area of 12 in.². If a smaller triangle is cut off the large ones so that all sides are 1/3 of the large triangle, what is the area?

The Circle

The circle's perimeter is called the *circumference*. $C = 2\pi r$, where C is the circumference, π is approximately 3.14, and r is the radius of the circle. The formula can also be written as $C = \pi d$, where d is the diameter of the circle.

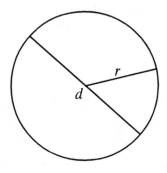

Figure 8.13

The *area* of the circle is $A = \pi r^2$.

Example:

Find the circumference and the area of a circle with a radius of 2 inches.

$C = 2\pi r = 2 \times 3.14 \times 2$ in. $= 12.56$ in.

$A = \pi r^2 = 3.14 \times (2)^2$ in.² $= 12.56$ in.²

Practice Problems:

8.30 The circumference of a circle is 20π. Find the radius.

8.31 The wheels of a toy car have a diameter of 1.5 cm. How far has the car moved when the wheels have rotated twice?

8.32 A semicircle (half circle) has a diameter of 3 inches. Find the area and perimeter of the semicircle.

8.33 A ring is formed from two concentric circles (circles with the same center). The diameter of the inner circle is 3/4 inch and the width of the ring is 1/4 inch. Find the plane area of the ring. See Figure 8.14.

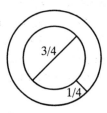

Figure 8.14

Solid Geometry

A *rectangular solid* looks like a box. It has 6 rectangular *faces*, 12 *edges*, and 8 *vertices* (corners). Figure 8.15 shows the box and Figure 8.16 shows the box flattened out.

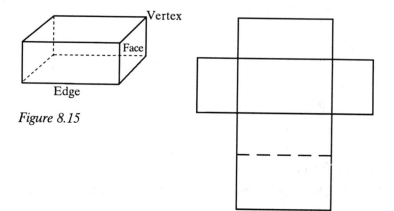

Figure 8.15

Figure 8.16

A *cylinder* looks like a can. Sometimes it has a top and a bottom like a can but sometimes it does not.

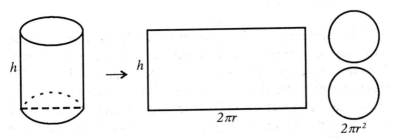

Figure 8.17

Area

The area of a rectangular solid equals the sum of the areas of the 6 faces.

Example:

A rectangular solid has the edges 3 in., 4 in., and 5 in. Find the total surface area.

There are 2 (3 × 4) rectangles, 2 (3 × 5) rectangles, and 2 (4 × 5) rectangles.

Total area: $2 \times 3 \times 4$ in.$^2 = 24$ in.^{2}

$2 \times 3 \times 5$ in.$^2 = 30$ in.^{2}

$2 \times 4 \times 5$ in.$^2 = 40$ in.^{2}

The surface area is $(24 + 30 + 40)$ in.$^2 = 94$ in.^{2}

Practice Problems:

8.34 A rectangular solid has edges of 3 in., 5 in., and 7 in. Determine the total surface area.

8.35 The edge of a cube is 10 cm. What is the total surface area?

8.36 If the length of each edge of the rectangular solid in problem 8.34 is to be increased by 20%, what is the surface area?

The surface area of a cylinder is as follows. If it has a top and a bottom, pretend that you remove both with a can opener. You now have two circles and a top/bottomless cylinder. Slit this cylinder vertically and flatten it. You have now obtained a rectangle with a width equal to the height of the cylinder and a length of $2\pi r$, where r is the radius of the circles that have been cut out. See Figure 8.17.

Example:

Find the surface area of a cylinder that has a height of 8 in. and two base areas with diameters of 4 in.

The base areas have each an area of $\pi(4/2)^2$ inches = $\pi(4)$ inches = 4π inches.

The area of the remaining rectangle is $\pi(4)(8)$ inches = 32π inches. Total area: 4π inches + 4π inches + 32π inches = 40π inches

Practice Problems:

8.37 A can is 10 inches high. The base area has a diameter of 6 inches. Find the total surface area.

Volume

The volume of a rectangular solid is width × length × height.

Example:

A rectangular solid has the edges 3 in., 4 in., and 5 in. Find the volume.

The width is 3 in., the length is 4 in., and the height is 5 in.

The volume is $3 \times 4 \times 5$ cubic inches (in.³) = 60 in.³.

Practice Problems:

8.38 Find the volume of a cube with an edge of 5 cm.

Example:

Find the volume of a cylinder that has a height of 8 in. and two base areas with a diameter of 4 in.

The volume of a cylinder is base × height = $\pi r^2 \times h$.

$r = 4/2$ inches = 2 inches. The radius is half of the diameter.

$h = 8$ inches

$V = \pi 2^2 \times 8$ in.^{2} = 32π in.^{2} or approximately 32×3.14 which is 100 in.^{2} (rounded).

Practice Problems:

8.39 A cylinder has a volume of 290 cm^3. The radius of the base cylinder is 5 cm. Determine the height of the cylinder.

8.40 The circumference of a cylinder is 4π in. and its height is 7 in. What is the volume of the cylinder?

Trigonometry

Trigonometry, the study of triangles, has become a major branch of mathematics. Here we are only looking at the very basic area of trigonometry, namely the trigonometry of right triangles.

In Figure 8.18, we have a right triangle with two angles equal to 45°. You can make your own 45° by folding the corner of a sheet of writing paper so that the sides (length and width) come together at a straight edge. With a ruler make two lines parallel to the side marked *leg*, as in Figure 8.18.

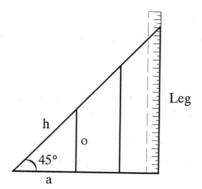

Figure 8.18

The leg next to the 45° angle is *adjacent*, the leg opposite the 45° angle is *opposite*, and the longest side is the *hypotenuse*. Measure the legs and the hypotenuses in all three triangles and fill in the following table.

Make a table:

Name	Adjacent	Opposite	Hypotenuse	o/h	a/h	o/a
Triangle 1						
Triangle 2						
Triangle 3						

Then, using your calculator, determine a/h, o/h, and o/a.

If you have a scientific calculator, find sin(e) 45°, cos(ine) 45°, and tan(gent) 45°. You usually have to key in the angle first: 45, cos. These are the trigonometric functions and you probably got very similar answers when you got them from the triangles.

In all 45° triangles these ratios are the same. All other angles have the same properties, so you can find the trigonometric functions by using a calculator or a table.

It is common to write SOHCAHTOA to help you to remember which legs in a right triangle give you which identity. SOH means sine = opposite over hypotenuse, CAH means cosine = adjacent over hypotenuse, and TOA means tangent = opposite over adjacent.

If you know the value of a certain trigonometric function, say tangent, you enter the value into your calculator and then the shift key.

For example, if the value of cosine is 0.5, find the angle. 0.5, shift, cos gives you 60. Is it true that cos 60° equals 0.5?

Example:

Find sine, cosine, and tangent of ∠*A* and of ∠*B*.

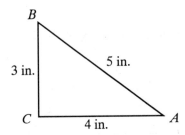

Figure 8.19

Determine a rule for the trigonometric functions of the nonright angles in right triangles.

$\sin A = 3/5 = 0.6$ $\sin B = 4/5 = 0.8$

$\cos A = 4/5 = 0.8$ $\cos B = 3/5 = 0.6$

$\tan A = 3/4$ $\tan B = 4/3$

Answer: The sine of one angle is equal to the cosine of the other angle. The tangents are reciprocals.

Practice Problems:

Use a calculator for these problems.

8.41 Find *a*) sin 23°, *b*) cos 62°, *c*) tan 55°.

8.42 Find the angle that has *a)* sin equal to 0.5 and *b)* tan equal to 1.

8.43 Find the value of *x* in Figure 8.20.

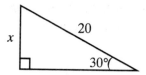

Figure 8.20

8.44 Find the value of α in Figure 8.21.

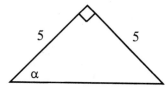

Figure 8.21

8.45 Find the length of the flagpole in Figure 8.22, if the shadow is 20 feet. Round to the nearest whole number.

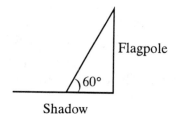

Figure 8.22

Analytic Geometry

This branch of mathematics is also called *coordinate geometry*, because points and equations are represented graphically in a coordinate plane.

The coordinate plane:

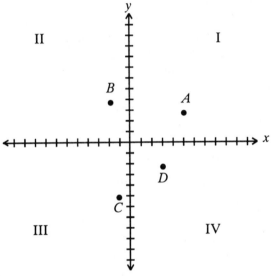

Figure 8.23

Two perpendicular number lines divide the plane into four parts, quadrants I, II, III, and IV. The horizontal number line is called the x-axis and the vertical number line is called the y-axis. Points are represented in the plane with two numbers (x,y) and are also called ordered pairs.

Example:

Plot the following points:

> *A*: (5,3)
>
> *B*: (−2,4)
>
> *C*: (−1, −5)
>
> *D*: (3,−2)

Use Figure 8.23 or make your own graph. Start at the point of intersection of the number lines, the origin, and move. A positive *x* tells you to move to the right and a positive *y* tells you to move up. A negative *x* goes to the left and a negative *y* goes down. If either *x* or *y* is zero, you get the message: "Stand still." You don't move in the zero direction.

Practice Problems:

8.46 Plot the following points:

> *A*: (0,5)
>
> *B*: (−3,0)
>
> *C*: (0,0)
>
> *D*: (6,−4)
>
> *E*: (−7,3)
>
> *F*: (−4,−4)

8.47 *a*) Plot (4,5) and (4,−3). Connect the points.
b) Plot (−2,4) and (2,4). Connect the points.
What conclusion can you draw when the *x*-numbers are equal? When the *y*-numbers are equal?

Example:

Go back to Figure 8.23 or your own graph and find the distance between the points in 8.47 *a* and between the points in 8.47 *b*. You can find these distances simply by counting the squares between them or take the difference between in *a* the *y*-numbers and in *b* the *x*-numbers.

Practice Problems:

8.48 Find the distance between *a*) points (5,9) and (5,5); *b*) points (5,5) and (3,5).

8.49 Find the distance between *a*) points (3,–2) and (3,–7); *b*) points (–4,–6) and (2,–6).

If we have two points in a coordinate system and want to find the distance between them, we can create a right triangle and use the Pythagorean Theorem. For example, the points in practice problem 8.49 form a right triangle with the sides 4 and 2 units. The distance between the points, that is, the hypotenuse in the right triangle, is

$$\sqrt{4^2 + 2^2} = \sqrt{16 + 4} = \sqrt{20} = 2\sqrt{5}$$

Example:

Find the distance between the points (4,–3) and (–2,5).

Make a right triangle on paper or in your mind: The horizontal leg is 5 – (–3) = 5 + 3 = 8. The vertical leg is 4 – (–2) = 4 + 2 = 6.

The distance is $\sqrt{8^2 + 6^2} = \sqrt{64 + 36} = \sqrt{100} = 10$

The distance is 10 units.

Practice Problems:

8.50   Find the distance between the points (0,–4) and (3,0).

8.51   Find the distance between the points (–2,6) and (4,3).

You can calculate the areas or perimeters of geometric figures by placing them in a coordinate system. There are many more things you can investigate with analytic geometry but here we are limited to a few.

Example:

Find the area of a triangle with vertices (–5,0), (3,0), and (0,4).

The base of the triangle is 3 – (–5) = 8 and the corresponding height is 4. Draw a figure to check if this is right! The area is 8(4)/2 square units = 16 square units.

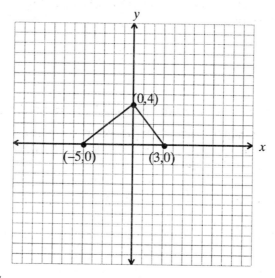

Figure 8.24

Practice Problems:

8.52 Find the area of a triangle with vertices (0,0), (10,0), and (5,3).

8.53 Find the area of a rectangle with vertices (5,6), (13,6), (5,2), and (13,2).

Example:

Find the area of a quadrilateral whose vertices are (2,3), (3,6), (10,4), and (4,2).

Mark the vertices on a coordinate system:

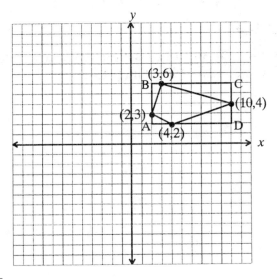

Figure 8.25

Draw horizontal lines through (4,2) and (3,6). Draw vertical lines through (2,3) and (10,4). Now we have one rectangle from we can subtract the areas of four right triangles.

The rectangle *ABCD* has vertices (2,2), (2,6), (10,6), and (10,2). The area of the rectangle is:
$(10 - 2)(6 - 2) = 8(4) = 32$ square units

The areas of the right triangles are:

Vertex *A*: $2(1)/2 = 1$ square unit

Vertex *B*: $1(3)/2 = 1.5$ square units

Vertex *C*: $7(2)/2 = 7$ square units

Vertex *D*: $6(2)/2 = 6$ square units

Total: 15.5 square units

The area of the quadrilateral is:
$(32 - 15.5)$ square units $= 16.5$ square units

Practice Problem:

8.54 Find the area of a quadrilateral whose vertices are (–2,2), (2,5), (8,1), and (–1,–2).

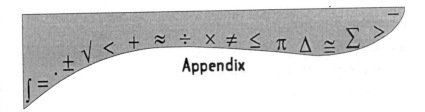

Review of Equations

The purpose of this appendix is to refresh your memory about equations and how they are solved.

We start with linear equations (they are called linear because you get a straight line when you plot the equation on a graph) with only one variable: x. You will also encounter equations with denominators. These can be proportions, which consist of two equal fractions, such as x over 4 equals 2 over 3, or non-proportions, when there are additional terms in the equation.

The next step is equations with two variables: x and y. These are called simultaneous equations, and we solve them to find out the numbers that x and y represent.

Finally, you learn to solve quadratic equations; they are called quadratic because they involve a square term, such as x^2. They are solved either by calculating a square root, if there are only squared terms in the equation, or by factoring.

Linear Equations With One Variable

When we solve equations, we want to find out what **number** x stands for. In other words, we want a new equation that states $x =$ a number.

Example:

Solve: $x + 5 = 7$.

Subtract 5 from both sides: $x = 2$

Check: $2 + 5 = 7$

Example:

Solve: $x - 5 = 7$.

Add 5 to both sides: $x = 12$

Check: $12 - 5 = 7$

Example:

Solve: $3x = 18$. *Note:* $3x$ means 3 times x.

Divide both sides by 3: $x = 6$

Check: $3(6) = 18$

Example:

Solve: $\dfrac{x}{3} = 18$. *Note:* This means x divided by 3.

Multiply both sides by 3: $3\left(\dfrac{x}{3}\right) = 3(18)$

$$x = 54$$

Check: $\dfrac{54}{3} = 18$

Example:

Solve: $22 = x + 15$. Here x is on the right side of the equation but that does not matter.

Subtract 15 from both sides.

$7 = x$ or $x = 7$

Check: $7 + 15 = 22$

Example:

Solve: $2x - 11 = 7$

Here we have two operations: multiplication and subtraction. The order of operations in solving equations is opposite to that in arithmetic. We "undo" what has been done before.

First add 11 to both sides: $\qquad 2x = 18$

Divide by 2: $\qquad x = 9$

Check: $2(9) - 11 = 18 - 11 = 7$

Example:

Solve: $\dfrac{x}{3} + 10 = 14$

Subtract 10 from both sides: $\qquad \dfrac{x}{3} = 4$

Multiply both sides by 3: $\qquad 3\left(\dfrac{x}{3}\right) = 3(4)$

$$x = 12$$

Check: $\dfrac{12}{3} + 10 = 4 + 10 = 14$

Example:

Solve: $\dfrac{2x}{3} + 5 = 15$

Subtract 5 from both sides: $\qquad \dfrac{2x}{3} = 10$

Multiply by 3: $\qquad 3\left(\dfrac{2x}{3}\right) = 3(10)$

$$2x = 30$$

$$x = 15$$

Check: $\dfrac{2(15)}{3} + 5 = 10 + 5 = 15$

Example:

$40 - 7x = 60 - 5x$

Should we add $7x$ or $5x$? It is easier if we add $7x$ to both sides, because then we avoid a minus sign before the x-term.

Add $7x$ to both sides:	$40 = 60 + 2x$
Subtract 60 from both sides:	$-20 = 2x$
Divide by 2:	$-10 = x$

Check: Left side $= 40 - 7(-10) = 40 + 70 = 110$ Multiplying two negatives → positive.

Right side: $60 - 5(-10) = 60 + 50 = 110$

Example:

Solve: $5 + (3x - 4) = 7 - (2x - 9)$

Remove the parentheses:	$5 + 3x - 4 = 7 - 2x + 9$
Simplify both sides:	$1 + 3x = 16 - 2x$
Add $2x$ to both sides:	$1 + 5x = 16$
Subtract 1 from both sides:	$5x = 15$
Divide by 5:	$x = 3$

Check: Left side $= 5 + (3(3) - 4) = 5 + 5 = 10$

Right side $= 7 - (2(3) - 9) = 7 - (6 - 9)$

$= 7 - (-3) = 7 + 3 = 10$

Equations With Denominators

Proportions

Example:

Solve: $\dfrac{x}{4} = \dfrac{3}{2}$

An equation that consists of two equal fractions only is called a proportion. It is usually solved by cross multiplication:

$$2x = 4(3)$$
$$2x = 12$$
$$x = 6$$

Check: $\dfrac{6}{4} = \dfrac{3}{2}$

Example:

Solve $\dfrac{5}{x} = \dfrac{10}{4}$

Cross multiply: $5(4) = 10x$
$$20 = 10x$$
$$x = 2$$

Check: $\dfrac{5}{2} = 2.5$ $\dfrac{10}{4} = 2.5$

Example:

Solve: $\dfrac{2x+3}{12} = \dfrac{x}{3}$

Cross multiply: $3(2x+3) = 12x$
$$6x + 9 = 12x$$
$$9 = 6x$$
$$x = \frac{3}{2} \text{ or } x = 1.5$$

Check:

$$\frac{2\left(\dfrac{3}{2}\right)+3}{12}=\frac{3+3}{12}=\frac{6}{12}=\frac{1}{2}$$

$$\frac{\dfrac{3}{2}}{3}=\frac{3}{2(3)}=\frac{1}{2}$$

Non-Proportion Equations

Example:

Solve: $\dfrac{x}{2}=9-\dfrac{x}{4}$

This equation is not a proportion because it contains three terms. A proportion consists of two equal fractions only.

To solve this equation we must find the least common denominator (LCD), which in this case is 4. Then we multiply all terms by 4.

$$4\left(\frac{x}{2}\right)=4(9)-4\left(\frac{x}{4}\right)$$

$$2x=36-x$$

Add x: $3x=36$

$$x=12$$

Check: $\dfrac{12}{2}=6$ $9-\dfrac{12}{4}=9-3=6$

Example:

Solve: $\dfrac{x}{3}-\dfrac{x}{5}=2$ LCD is 15.

$$15\left(\frac{x}{3}\right) - 15\left(\frac{x}{5}\right) = 15(2)$$

$$5x - 3x = 30$$

$$2x = 30$$

$$x = 15$$

Check: $\dfrac{15}{3} - \dfrac{15}{5} = 5 - 3 = 2$

Simultaneous Equations

When solving word problems, we often need two variables: x and y. To solve for both, we need two equations.

Example:

Solve for x and y:

$$x + y = 6$$
$$x - y = 2$$

Add the equations:
$$2x = 8$$
$$x = 4$$
$$4 + y = 6$$
$$y = 2$$

Check: $4 + 2 = 6$ $\qquad$ $4 - 2 = 2$

Example:

Solve for x and y:

$5x - 2y = 5$ Multiply by 3.	$15x - 6y = 15$
$x + 3y = 18$ Multiply by 2.	$2x + 6y = 36$
Add:	$17x = 51$
	$x = 3$

Replace x in one of the original equations:

$3 + 3y = 18$

$\quad 3y = 15$

$\quad\; y = 5$

Check: $5\,(3) - 2(5) = 15 - 10 = 5$

$\qquad\quad\; 3 + 3(5) = 3 + 15 = 18$

Alternate solution:

Multiply the second equation by 5: $5x + 15y = 90$

Subtract the first equation from the second: $5x - 2y = 5$

$\qquad\qquad\qquad\qquad\qquad\qquad\qquad 17y = 85$

$\qquad\qquad\qquad\qquad\qquad\qquad\qquad\quad y = 5$

$\qquad\qquad\qquad\qquad\qquad\quad 5x - 2(5) = 5$

$\qquad\qquad\qquad\qquad\qquad\qquad\quad 5x = 15$

$\qquad\qquad\qquad\qquad\qquad\qquad\quad\; x = 3$

Example:

Solve for x and y:

$y = x + 3$

$2x + y = 9$

Here we could rewrite the first equation as $-x + y = 3$ but it is easier to substitute the first equation into the second:

$2x + x + 3 = 9$

$\qquad 3x = 6$

$\qquad\; x = 2$

$\qquad\; y = 2 + 3 = 5$

Check: $2(2) + 5 = 4 + 5 = 9$

Quadratic Equations

Sometimes our equations contain an x^2-term. If there is no x-term, we can solve the equation by taking the square root of both sides.

Example:

Solve $x^2 = 25$

$$x = \pm\sqrt{25}$$
$$x = \pm5$$

The x-term can be 5 or –5. However, in a word problem we must be sure that we can use the negative answer. For example, if the problem stated: The square of a whole number is 25, we have to reject –5, because the whole numbers are 0, 1, 2, 3,.... but if the problem stated: The square of an integer is 25, we give both 5 and –5 as answers.

Often quadratic equations also contain an x-term. Such equations have to be solved by factoring or formula. We are not going to cover the formula here.

In order to factor, we need all terms on one side of the equation. It is usually easiest to have them to the left. Then we look at the constant term, in the following example +5. We need two integers whose product is +5. Here we have to guess –1 and –5. Then we add the two numbers we found. Do they add up to the coefficient of x (that is, the number before x)? In this case: $-1 + -5 = -6$.

Example:

Solve: $x^2 - 6x + 5 = 0$.

Factor: $(x - 5)(x - 1) = 0$

This is a true statement, if each factor equals zero.

We set each factor $= 0$.

$$x - 5 = 0 \qquad x - 1 = 0$$
$$x = 5 \qquad x = 1$$

Answer: x can either be 5 or 1.

Example

Solve: $x^2 - 6x + 9 = 0$

Factor: $(x - 3)(x - 3) = 0$

$$x - 3 = 0 \qquad x - 3 = 0$$
$$x = 3 \qquad x = 3$$

The two solutions are both equal to 3.

Example:

Solve $x^2 + x - 2 = 0$

Factor: $(x - 1)(x + 2) = 0$.

$$x - 1 = 0 \qquad x + 2 = 0$$
$$x = 1 \qquad x = -2$$

Check: If $x = 1$ then $1^2 + 1 - 2 = 0$

If $x = -2$ then $(-2)^2 + (-2) - 2 = 4 - 2 - 2 = 0$

Answer: $x = 1$ or $= -2$.

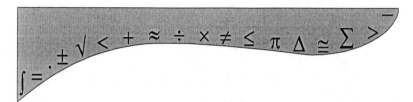

Answers to Practice Problems

Chapter 1

1.1　$x + x + 10 = 20$
$$x = 5$$
$$x + 10 = 15$$
The pieces are 5 and 15 in.

1.2　The pieces are 3, 4, and 5 ft.

1.3　The pieces are 3 and 6 ft.

1.4　The pieces are called x, $2x$, and $2x + 10$.
They are 14, 28, and 38 in.

1.5　$x + x + 50\%x = 275$
$$2.5x = 275$$
$$x = 110$$
$$1.5x = 165$$
The ropes are 110 and 165 yards.

1.6　Elsa is 14 years old and Thor is 21 years old.

1.7　The grandmother is 84 years old.

1.8   If Jessica is x, the mother $x + 28$, and the grandmother
       $2(x + 28)$
       $x + x + 28 + 2(x + 28) = 100$
       $x = 4$
       The grandmother is $2(4 + 28) = 64$ years old.

1.9 $E = 3L$ $\rightarrow 2E - 3L = 0$
       $E + L = 65$ $\rightarrow 3E + 3L = 195$
                                 $5E = 195$
                                 $E = 39$
                                 $L = 65 - 39 = 26$
       Answer: Eric is 39 years old and Lucas is 26 years old.

1.10   Michael is 12 years old.

1.11   The number is 26.

1.12   The sun rose at 7:48 a.m.

1.13   The water level was 67.2 ft.

1.14   The population was 503.

1.15   Joan weighted 133 lbs.

1.16 $-6 > -10$

1.17 $x \geq 18$

1.18 $x \geq 8$

1.19   7, 9, 11, 13, 15

1.20 $-6, -4, -2$

1.21 $-2, -1, 0, 1, 2$

1.22   The first integer is 10.

1.23   The product of 6 and 8 is 48.

1.24   Call the integers $x, x + 2$ and $x + 4$
       $6x = 5(x + 2)$
       $x = 10$
       The largest integer is 14.

1.24   The integers are 45, 46, 47, and 48.

1.26   Call the integers $x, x + 2, x + 4$.
$3x = x + 2 + x + 4$
The integers are 6, 8, and 10.

1.27   The first odd integer is 7.

1.28   The even integers are 4, 6, and 8.

1.29   Call the integers x and $x + 1$

$$x(x + 1) = 20$$
$$x^2 + x - 20 = 0$$
$$(x + 5)(x - 4) = 0$$
$$x + 5 = 0$$
$$x = -5 \qquad \text{Reject: } x \text{ is a positive integer}$$
$$x - 4 = 0$$
$$x = 4$$
$$x + 1 = 5$$

The consecutive positive integers are 4 and 5.

1.30 $x(x + 4) = 5$
$x = -5 \qquad\qquad x = 1$
$x + 4 = -1 \qquad x + 4 = 5$
The numbers are –5 and –1 or 1 and 5.

1.31   The integers are –5 and –3 or 3 and 5.

1.32   The numbers are 9 and 27 or –9 and –27.

1.33   The numbers are –22 and –20 or 20 and 22.

Chapter 2

2.1   1/300

2.2   9/200 or 0.045

2.3   0.00006

2.4   0.017

2.5   4

2.6   300

2.7   0.1875

2.8   525
2.9   0.4
2.10 12%
2.11 0.56%
2.12 1200%
2.13 80%
2.14 66 2/3% or 66.67%
2.15 50,000%
2.16 25%
2.17 6350
2.18 1050
2.19 80%
2.20 $700
2.21 $120
2.22 350
2.23 $80
2.24 $3000
2.25 $11.50
2.26 2.88%
2.27 50%
2.28 15.5%
2.29 16.4%
2.30 $265,000
2.31 $31.50
2.32 5%
2.33 5%
2.34 $127.50
2.35 $200.60
2.36 $100

2.37 $180

2.38   15.33%

2.39   90¢

2.40   Both cost the same: $200.

2.41   You pay $42; the order does not matter.

2.42 $153

2.43 $89.76

2.44   12.2%

2.45   653

2.46   *a*) 2%   *b*) 2 years

2.47   6%

2.48 $501.88

2.49 $502.85

2.50 $21,333.33

2.51 $7.67

2.52 $1000 + 1000(10\%)(1/12) = 1008.33$
       $1008.33 + 1008.33(10\%)(1/12) = 1016.73$
       $1016.73 + 1016.73(10\%)(1/12) = 1025.20$
       $1025.20 - 1000 = 25.20$
       Answer: The interest was $25.20.

2.53 $500(1 + 2\% \div 12)^{12} = 510.09$
       Answer: $510.09 – $500 = $10.09

2.54 $700(1 + 1.5\% \div 365)^{365} = 710.58$
       Answer: $710.58

2.55 $12,000(1 + 12\%)^4 = 18,882.23$
       $18,882.33 - 12,000 = 6,882.23$
       Answer: Barbara owes $6,882.23.

2.56 $6800(1 + x\%)^5 = 10,000$
       $(1 + x\%)^5 = 1.4706$
       Replace *x* by 6, 7, 8, 9 until you get a true statement.
       $(1 + 8\%)^5 = 1.4693$, which is approximately 1.47.

2.57   Day-to-Day: $1500(1+0.5\%)^3 = 1522.61$
Interest $= 1522.61 - 1500 = 22.61$
Credit Union: $1500(1 + 2.72\% \div 4)^{12} = 1627.08$
Interest $= 1627.08 - 1500 = 127.08$
$127.08 - 22.61 = 104.47$
Answer: The Credit Union pays $104.47 more.

2.58   Bought: $100 \times \$10 = \$1000$
Sold: $100 \times \$12.05 = \$1205$
Dividend: $4 \times \$1.25 = \$5.00$
Earnings: $\$1205 + \$5.00 - \$1000 = \$210$
Jim's profit was $210.

2.59   a) $7 \times 10 \times 15 = \$1050$
b) $10 \times 100 = \$1000$
c) $15/50 = 30\%$

2.60   Cost + mark-up $70
Selling price: 0.85 of $70 = $59.50
Profit: $9.50

2.61   Cost: $60/1.2 = 50$
Selling price 0.8 of $50 = 40$
The selling piece is $40.

Chapter 3

3.1   If Elsa is x, then Thor is $x + 7$.
Equation: $x + x + 7 = 35$
Elsa is 14 years old and Thor is 21 years old.

3.2   I am x and my father is $x + 41$.
$x + 41 - 8 = 3(x + 5)$
$x = 9$ I am 9 years old.

3.3

	Today	5 years ago
Lyn	x	$x - 5$
Father	$4x$	$4x - 5$

$4x - 5 = 7(x - 5)$
$x = 10$ Lyn is 10 years old.

3.4 $E + C + F = 61$
$E = C + 5$
$F = 6C$
$C + 5 + C + 6C = 61$
$8C = 56$
$C = 7$ Carl is 7 years old.

3.5 In x years: $6 + x + 3 + x + 1 + x = 0.8(40 + x)$
$3x + 10 = 32 + 0.8x$
$\quad 2.2x = 22$
$\qquad x = 10$ In 10 years.

3.6 Eva is 44 years old.

3.7 Susan is 12 years old and Jack is 15 years old.

3.8 If Carla is x, then Glenn is $x + 6$ years old.
$2(x + 6) + x = 57$
$x = 15 \quad x + 6 = 21$
Carla is 15 years old and Glenn is 21 years old.

3.9 Mary is 18 years old and Chris is 11 years old.

3.10 $5R - 3S = S$
$R = S - 2$
Sig is 10 years old and Ray is 8 years old.

3.11 $A + V = 15$
$V = 1/2\,A \rightarrow 2V = A$
$2V + V = 15$
$\quad 3V = 15$
$\qquad V = 5$ Victor is 5 and Adam is 10 years old.

3.12 Ina, Mina, and Mo are 24, 36, and 42 years old.

3.13 If Mark is x and Mindy is y,
then $x + y = 84$ and $3x = 4y$.
Mark is 48 years old and Mindy is 36 years old.

3.14 Jim and Jon are $3x$ and $7x$, respectively.
$4(3x) = 7x + 40$
$\quad x = 8$
Jim is $3(8) = 24$ years old.
Jon is $7(8) = 56$ years old.

3.15   David is 8 years old.

3.16   The mother is now 38 years old.

3.17   Ellen is 10 years old.

3.18   John is 9 and Ed is 17 years old.

3.19   Ginger is 8 years old.

3.20   Bev is x and Ron is $x + 6$.
$$2(x + 10) + 1 = 3(x - 3)$$
$$x = 30$$
Bev is 30 and Ron is 36 years old.

3.21   Phil is 27 years old. Reject –27 because an age cannot be negative.

3.22   Ronald is 20 and Liz is 22 years old. Reject the negative answers.

3.23   The older daughter is 23 years old.

3.24   The first perfect square is $45^2 = 2025$. The mathematician will be 45 years old in 2025. In 2006 she will be $2025 - 2006 = 19$ years younger or $45 - 19 = 26$. The mathematician will be 26 years old in 2006.

3.25 $S = L + 5$
$L^2 + 2S = 58$
$L^2 + 2L - 48 = 0$
$(L + 8)(L - 6) = 0$
$L = -8$ Reject
$L = 6$ Lucy is 6 years old.

Chapter 4

4.1   40 $3.85 stamps and 10 80¢ stamps

4.2   20 nickels and 40 dimes

4.3   32 nickels and 14 dimes

4.4   25 quarters and 2 dimes

4.5 $20\%x + 50\%(80) = (80 + x)45\%$
          $20x + 50(80) = (80 + x)45$
                      $x = 16$
16 liters of the 20% solution

4.6   60 oz   20%   60(20%)
                 x oz   100%   100x% x
             60 + x   40%   (60 + x)40%

$60(20\%) + 100\% x = (60 + x)40\%$
Multiply by 100:
          $60(20) + 100x = (60 + x)40$
                           $x = 20$
Add 20 oz. pure acid.

4.7   The punch has 19% (rounded) of pomegranate juice.

4.8   5 liters of water

4.9   3 liters of water

4.10   67 quarts (rounded)

4.11   Add 3.5 liters of water.

4.12   60 g of nickel

4.13   300 g   14-c   300(14)
                 x g   24-c   24x
             300 + x   18-c   18(300 + x)

$300(14) + 24x = 18(300 + x)$
                   $x = 200$
Add 200 g pure gold.

4.14   A chicken has 2 legs and a sheep has 4 legs.
Total number of heads: $x + y = 22$
Total number of legs: $2x + 4y = 58$
Solve the system of equations:
$x = 15$ and $y = 7$
There are 15 chickens and 7 sheep.

4.15   Selma has 60 $10 bills.

4.16   There were 7 $20 bills.

4.17 $2x + 3(x - 50) = 1450$
$$x = 320$$
An adult ticket was $320.

4.18 $5\,g + 8\,c = 170$
$7\,g + 4\,c = 130$
$$g = 10$$
$$c = 15$$
The gum cost 10¢ and the chocolate cost 15¢.

4.19   40 regular hours + 7 hours overtime = $390
42 regular hours + 8 hours overtime = $416
Kelly's overtime rate is $10 per hour.

4.20

	Amount	Percent	Interest
CD	x	1.20%	$1.2x\%$
Bonds	$3000 - x$	3%	$3\%(3000 - x)$
			$72

$1.2x\% + 3\%(3000 - x) = 72$
Multiply both sides by 100:
$$1.2x + 9000 - 3x = 7200$$
$$x = 1000$$
$3000 - 1000 = 2000$
Paul invested $2000 in bonds.

Chapter 5

5.1   5 min. : 60 min. = 1:12

5.2 $3x + 4x = 21$
$$x = 3$$
$$3x = 9$$
$$4x = 12$$
The pieces are 9 and 12 in.

5.3   3 women + 7 men = 10 total
7/10 of 4680 men = 3276
3/10 of 4680 women = 1404
There were 3276 men and 1404 women.

5.4   16 students + 1 professor = 17 total
1/17 of 3400 = 200
200 professors at the game.

5.5   8 women, 20 people, 12 men
Men: women = 12:8 = 3:2
The ratio of men to women is 3:2.

5.6   462 miles

5.7 $\dfrac{48}{6} = \dfrac{256}{x}$
$x = 32$

The cost is 32¢.

5.8   9 in.

5.9   20

5.10   100 times

5.11 $\dfrac{x}{6} = \dfrac{48}{8}$
$x = 36$

The flagpole is 36 ft. tall.

5.12   42 in.

5.13   3/4 lb.

5.14   18 cups

5.15   1/2 pint

5.16   4.5 ft.

5.17   13 5/6 years

5.18   80 servings

5.19   80 square yards

5.20   1760 yards

5.21   5 lbs. peaches
2 2/9 cups apricot jam

5.22   1296 in.^{2}

5.23

Unit	Teaspoons	Tablespoons	Cups	Pints	Quarts
1 tsp.			1/48	1/96	1/192
1 Tbsp.					1/64
1 pint	96	32			

5.24   25 Tbsp. = 25/16 cups = 1 9/16 cups

5.25   7 cans

5.26   14 Tbsp. are left.

5.27   3 1/8 pints

5.28   4500 g

5.29   380 cm

5.30   80 m

5.31   3.890 kg

5.32   1.5 dm

5.33   1.50 cm^2

5.34   5 mm^2

5.35   75 cm^2

5.36   3000 dm^3

5.37   5000 cm^3

5.38   50 cL

5.39   3 dm

5.40   20 small units

5.41   0.15 m × 0.12 m = 15 cm × 12 cm

5.42   100 × 125 mm = 12500 mm= 12.5 m

5.43   3 × 200 mm = 600 mm = 60 cm

5.44   1 cup = 1/4 quarts = 0.946 L / 4 = 9.46 dL/4 = 2.4 dL
       (rounded)

5.45   3.4 kg = 7.48 lb.
       51 cm = 20 in.

5.46   3.57 oz.

5.47   29.9 in.

5.48   11.6¢

5.49   5mm/sec.

5.50   40 km/hr.

5.51 $1.17

5.52   1640 ft.

5.53   *a*) –40   *b*) 32   *c*) 50   *d*) 212

5.54   *a*) –40   *b*) –23   *c*) 0   *d*) 100

5.55   The babysitter thought that the temperature was measured in Celsius degrees, not in Fahrenheit. 100 °C = 212 °F, the temperature of boiling water!

Chapter 6

6.1   4 mph

6.2   They meet after 2 hours, 30 miles from Carl's house.

6.3   The speeds are 50 and 60 mph.

6.4

	R	T	D
Up	3.3/5 mph = 0.66 mph	5 hr.	3.3 mi.
Down	3.3/1.5 mph = 2.2 mph	1 1/2 hr.	3.3 mi.

The difference between the rates is 1.54 mph.

6.5   Total trip: 400 miles
200/45 hr. = 4.4 hr.
200/55 hr. = 3.6 hr.
Total time: 8 hours
Average speed: 400/8 mph = 50 mph

6.6   The car went on the dirt road for 1 hour.

6.7   They met after 1 2/7 hour or 1 hour 17 min.
The time is then 11:57.

6.8

	R	T	D
Old	r	d/r	d
New	1.2 r	9/12 d/r	0.9d

The new time is 3/4 of the old time, or 25% shorter.

6.9   Rate in still water: 6 mph
Rate of current: 2 mph

6.10   Speed in still air: 525 mph
Speed of wind: 75 mph

6.11   2 hours

6.12   40 minutes

6.13 $\dfrac{2}{3} + \dfrac{2}{x} = 1$

$x = 6$

It takes Jonas 6 hours to do the job alone.

6.14

	R	T	W
A + B	1/12	8 hr.	8/12
C	1/x	8 hr.	8/x
A + B + C	1/8	8 hr.	1

8/12 + 8/x = 1
$x = 24$
Alternate solution:
1/12 + 1/x = 1/8
2x + 24 = 3x
$x = 24$
It would take Carl 24 hours to do the job alone.

6.15   120 minutes

6.16   3 R 17 cars 10 min.
1 R 17/(30) cars/min.
14 R 14(17)/30 cars/min.
14 R   14(17)45/30 cars/min.
14 robots can assemble 357 cars.

6.17   20 minutes

6.18   3 hours

6.19 $x/12 - x/30 = 1$
           $x = 20$
       It takes 20 minutes.

Chapter 7

7.1   Mean   *a)* 5   *b)* 8 (rounded)   *c)* –4 (rounded)
       Median   6   8   –4
       Mode   1 and 7   8   –5
       Range   8   4   7

7.2 $\dfrac{12 \times 25 + 13 \times 20}{26} = \dfrac{560}{25} = 22.4$

7.3   23 (rounded)

7.4   24 (rounded)

7.5   34.5

7.6   The shape of the bar graph and frequency polygon will depend on the frequency of heads you obtain.

7.7   Mean   38.3
       Median   37
       Mode   none
       Range   52

7.8   Mean   7.8
       Median   8
       Mode   8
       Range   4

7.9 | Frequency | Distribution |
       | --- | --- |
       | 0 | 2 |
       | 1 | 5 |
       | 2 | 6 |
       | 3 | 3 |
       | 4 | 2 |
       | 5 | 1 |
       | 6 | 1 |

Mean   2.25
Median   2
Range   6

7.10 Mean   5.7
Median   5.5
Mode   5

7.11 a) Salaries for teachers $\dfrac{52\% \times 4.3}{709}$ mil. = \$3154

b) Upkeep of buildings $\dfrac{26\% \times 4.3}{709}$ mil. = \$1577

7.12 Men + Women over 65: 1,543,332
Total population: 8,837,496

$$\dfrac{1,543,332}{8,837,496} \times 360° = 63°$$

7.13 a) \$125,000   b) \$75,000

7.14 a) $\dfrac{15}{60} = \dfrac{1}{4}$ b) $\dfrac{20}{60} = \dfrac{1}{3}$ c) $\dfrac{5}{60} = \dfrac{1}{12}$

7.15 6/10 = 3/5, or 0.6

7.16 Total number of minutes in 24 hours: 1440 minutes.
Equal digits: 1.11, 2.22, 3.33, 4.44, 5.55, 11.11.
11.11 occurs once, the others twice during 24 hours.
Probability: 11/1440 = 0.0076 or 0.76%

7.17 1/52

7.18 6/52 = 3/26, or 11.53%

7.19 a) 1/6   b) 3/6 = 1/2   c) 2/6 = 1/3   d) 0

7.20 a) 1/8   b) 3/8   c) 1/8

7.21 a) $\dfrac{25}{84}$ b) $\dfrac{40}{84} = \dfrac{10}{21}$ c) $\dfrac{29}{84}$

7.22 3:3 = 1:1

7.23 $3:3 = 1:1$

7.24 $\dfrac{1}{6} \times \dfrac{1}{6} = \dfrac{1}{36}$

7.25 $30\% \times 30\% = 9\%$

7.26 $\dfrac{1}{6} \times \dfrac{3}{6} = \dfrac{1}{12}$

7.27 $\dfrac{1}{2} \times \dfrac{1}{6} = \dfrac{1}{12}$

7.28 $a) \dfrac{2}{9} \times \dfrac{2}{9} = \dfrac{4}{81}$ $b) \dfrac{2}{9} \times \dfrac{1}{8} = \dfrac{1}{36}$

7.29 $\dfrac{7}{25} + \dfrac{10}{25} = \dfrac{17}{25}$

7.30 $\dfrac{4}{52} + \dfrac{4}{52} = \dfrac{8}{52} = \dfrac{2}{13}$

7.31 $\dfrac{1}{6} + \dfrac{3}{6} = \dfrac{4}{6} = \dfrac{2}{3}$

7.32 $\dfrac{6}{36} + \dfrac{3}{36} - \dfrac{1}{36} = \dfrac{8}{36} = \dfrac{2}{9}$

7.33 $\dfrac{3}{6} + \dfrac{1}{6} + \dfrac{1}{6} - \dfrac{1}{6} = \dfrac{4}{6} = \dfrac{2}{3}$

7.34 $\dfrac{1}{6}\left(\dfrac{1}{6} + \dfrac{1}{6}\right) = \dfrac{1}{6} \times \dfrac{1}{3} = \dfrac{1}{18}$

7.35 $534 \times 216 = 115{,}344$ dates

7.36 $(3)(3)(3) = 27$

7.37 $5 \times 4 \times 3 \times 2 \times 1 = 120$ ways

7.38   (2)(3)(2)(1) = 12

7.39   (3)(2)(1)(1) = 6

7.40   CIRCLE has 6 letters but 2 are the same.
       The number of arrangements is 6!/2! = 360

7.41   {2, 4, 6,...}

7.42   *a*) the empty set { } or $\emptyset$
       b) {1, 2, 3, 4, 5, 6, 8]

7.43   36 students are taking German.

7.44   14 people could both sing and dance.

7.45   20 cars have both an airbag and a car phone.

Chapter 8

8.1   Complementary   Supplementary
       a)   45°   135°
       b)   60°   150°
       c)   0°   90°
       d)   23°   113°
       e)   15°   105°

8.2   w = 7 cm   l = 14 cm

8.3   w = 7 units   l = 13 units

8.4   w = 13 cm   l = 18 cm

8.5   20 in.

8.6   l = 4.5 in.   w = 3 in.

8.7   14.4 in.

8.8   13 cm, 26 cm, and 21 cm

8.9   36 in.

8.10   20.25 in.²

8.11 $x^2 = (x + 2)(x - 1)$
       $x = 2$
       The area is 4 in.²

8.12 $\quad 2x + 2y = 24$

$\qquad 3x \times 2y = 160 + xy$

$\qquad\qquad xy = 32$

$\qquad\qquad\quad y = 12 - x$

$\qquad x(12 - x) = 32$

$x^2 - 12x + 32 = 0$

$(x - 8)(x - 4) = 32$

$\quad x = 8 \qquad\qquad x = 4$

$\quad y = 4 \qquad\qquad y = 8$

The sides are 4 in. and 8 in.

8.13 Area of rectangle: $11.2 \times 6 = 67.2$

$$\frac{67.2 - 64}{64} = 5\%$$

The area is 5% larger.

8.14 8 in.

8.15 $\dfrac{5b}{2} = 30$

$\qquad b = 12$ inches

8.16 10 units

8.17 25 units

8.18 The area is 6 in.^{2}

$\qquad 3b/2 = 6$

$\qquad\quad b = 4$

This is a right triangle with the legs 3 in. and 4 in.

8.19 $\sqrt{65 - 49} = \sqrt{16} = 4$

The missing leg is 4 units.

8.20 $\sqrt{20^2 - 8^2} = \sqrt{336} = 18.3$

The ladder is 18.3 feet up on the wall.

8.21 $A = 80° \qquad\qquad B = 40° \qquad\qquad C = 60°$

8.22 $A = 90°$

8.23 $3x + 5x + 52 = 180$
$$x = 16$$
$$3(16) = 48$$
$$5(16) = 80$$
The angles are 48° and 80°.

8.24 $A + B = 156$
$A = C$
$A + B + C = 180$
Angle $B = 132°$

8.25 $x = 10$

8.26 $x = 4.5$

8.27 The sides are 15 and 18 cm.

8.28 The tree is 36 ft.

8.29 $12/9 = 4/3$
The area is 1 1/3 in.²

8.30 $r = 10$

8.31 $\pi \cdot 1.5 \cdot 2 = 3\pi$ cm or about 9.4 cm

8.32 $A = \dfrac{1}{2}\pi\left(\dfrac{3}{2}\right)^2 = \dfrac{9}{8}\pi$ in.² $= 3.5$ in.²

$P = \dfrac{1}{2}3\pi + 3 = 1.5\pi + 3$ in. $= 7.7$ in.

The area is approximately 3.5 in.² and the perimeter is 7.7 inches.

8.33 $R = 5/8$ in.
$r = 3/8$ in.
The area of the ring is $R^2 - r^2$.
The area is $\pi/4$ sq. in. or 0.79 in.².

8.34 142 in.²

8.35 600 in.²

8.36 204.48 in.²

8.37 $60p + 2(9p) = 78p$
The total surface area is 78π in.^{2}, or
245 in.^{2}.

8.38 125 cm^3

8.39 $25\pi h = 290$
$h = 3.7$ cm (approximately)

8.40 $\pi\,(4)(7) = 28\pi$ or 88
88 in.^{3}

8.41 *a*) 0.39   *b*) 0.47 *c*) 1.43

8.42 *a*) 30°   *b*) 45°

8.43 $\sin 30° = x/20$
$x = 10$

8.44 $\tan \alpha = 5/5$
$\alpha = 45$
The angle is 45°

8.45 $\tan 60 = x/20$
$x = 34.6$
The length of the flagpole is 34.6 ft.

8.46 *A* is on the $+y$-axis
B is on the $-x$-axis
C is the origin where the axes cross
D is in the IV quadrant
E is in the II quadrant
F is in the III quadrant

8.47 When the *x*-numbers are equal, the line between points
is vertical.
When the *y*-numbers are equal, the line is horizontal.

8.48 *a*) 4   *b*) 2

8.49 *a*) 5   *b*) 6

8.50 $\sqrt{9+16} = 5$

8.51 $\sqrt{36+9} = \sqrt{45}$

8.52 $10 \times 3 / 2 = 15$
       15 square units

8.53 $(13 - 5)(6 - 2) = 8(4) = 32$
       32 square units

8.54   Use the example as a model.
       The rectangle is 70.
       The triangles are 2, 6, 12, 13.5.
       The polygon is 36.5 square units.

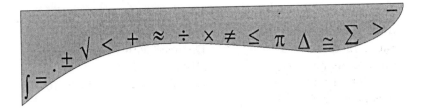

Index

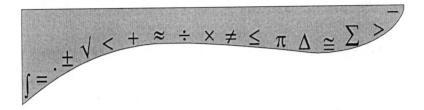

About the Author

BRITA IMMERGUT taught mathematics for 30 years in middle schools, high schools, and colleges. She was a professor of mathematics at LaGuardia Community College of the City University of New York. She has conducted workshops and taught courses for math-anxious adults at schools and organizations. Professor Immergut received an M.S. in mathematics, physics, and chemistry from Uppsala University in Sweden and an Ed.D. in mathematics education from Teachers College, Columbia University. She is the author of *How to Help Your Child Excel in Math* (Career Press) and the co-author of two textbooks for adults: *Arithmetic and Algebra...Again* and *An Introduction to Algebra: A Workbook for Reading, Writing and Thinking about Mathematics*.

TELL THE WORLD
THIS BOOK WAS

Good	Bad	So-so